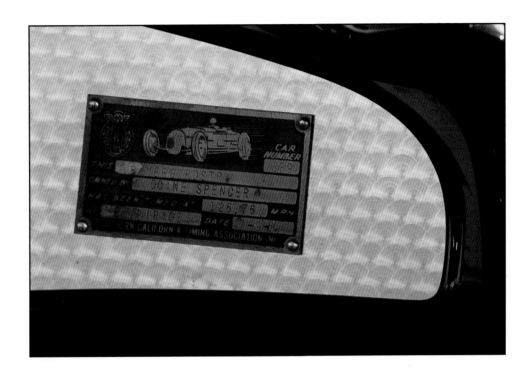

HOT ROD
Milestones

Dain Gingerelli

MBI Publishing Company

Dedication

In memory of Tom McMullen. He opened the door for many aspiring writers to join the hot rod magazine business. Mine came in 1971. No doubt, Tom could be jaded and brash, but above all else, he was a hot rodder.

First published in 1999 by MBI Publishing Company, 729 Prospect Avenue, PO Box 1, Osceola, WI 54020-0001 USA

© Dain Gingerelli, 1999

All rights reserved. With the exception of quoting brief passages for the purpose of review no part of this publication may be reproduced without prior written permission from the Publisher.

The information in this book is true and complete to the best of our knowledge. All recommendations are made without any guarantee on the part of the author or Publisher, who also disclaim any liability incurred in connection with the use of this data or specific details.

We recognize that some words, model names and designations, for example, mentioned herein are the property of the trademark holder. We use them for identification purposes only. This is not an official publication.

MBI Publishing Company books are also available at discounts in bulk quantity for industrial or sales-promotional use. For details write to Special Sales Manager at Motorbooks International Wholesalers & Distributors, 729 Prospect Avenue, Osceola, WI 54020-0001 USA.

Library of Congress Cataloging-in-Publication Data

Gingerelli, Dain.
  Hot rod milestones / Dain Gingerelli.
    p.   cm.
  Includes index.
  ISBN 0-7603-0637-0 (pbk. : alk. paper)
  1. Hot rods—History. I. Title.
TL236.3.G56   1999
629.228'6—dc21   99-20205

On the front cover: A hot rod must be truly extraordinary to win the America's Most Beautiful Roadster award. The *Emperor* filled that prerequisite in 1960 when Chuck Krikorian's car topped all others at the National Roadster Show in Oakland, California. The car remains in its original condition today, right down to the same paint, interior, and engine.

On the frontispiece: One way to boast about a hot rod's heritage is to put its dry lakes timing tag on the dashboard. Despite a top speed of 126 miles per hour at El Mirage, the Doane Spencer-built 1932 Ford highboy is most remembered for its timeless beauty. The car was originally completed shortly after World War II and was recently restored by the SO-CAL Speed Shop for Bruce Meyer.

On the title page: The first car to win the America's Most Beautiful Roadster award was Bill NeiKamp's 1929 Ford. Jim "Jake" Jacobs acquired the car in 1969, restoring it to its former glory. I photographed the car for this book on the Petersen Automotive Museum parking deck.

On the back cover: Ed Iskenderian's 1924 Model T roadster was among the first 12 cover cars in *Hot Rod* magazine's first year of publication. It appeared on the June 1948 cover, and was featured as the "Hot Rod of the Month."

Edited by Keith C. Mathiowetz
Designed by Rebecca Allen

Printed in Hong Kong

CONTENTS

	Acknowledgments	**7**
	Introduction	**11**
Chapter 1	**Downey Brothers' 1926 Ford Roadster**	**19**
Chapter 2	**John Athan's 1929 Ford Roadster**	**25**
Chapter 3	**Ed Iskenderian's 1924 Ford Roadster**	**31**
Chapter 4	**Doane Spencer's 1932 Ford Roadster**	**37**
Chapter 5	**Bill NeiKamp's 1929 Ford Roadster**	**43**
Chapter 6	**Dick Kraft's 1924 Ford Model T**	**49**
Chapter 7	**Dick Williams' 1927 Ford Roadster**	**55**
Chapter 8	**Chuck Krikorian's 1929 Ford Roadster**	**61**
Chapter 9	**Mike Haas' 1923 Ford**	**67**
Chapter 10	**Dan Woods' 1917 Ford Roadster**	**73**
Chapter 11	**Lil' John Buttera's 1926 Ford Sedan**	**79**
Chapter 12	**Tom McMullen's 1932 Ford Roadster**	**85**
Chapter 13	**Jim "Jake" Jacobs' 1928 Ford Touring**	**91**
	Index	**96**

It's all smiles for the camera! During a time-out with Dick Kraft's *The Bug*, I asked Richard Campos if he'd take a snapshot of me with the car. I'm mimicking Krafty Dick's famous pose behind the wheel at the 1950 Santa Ana Drags.

Acknowledgments

Some people view the task of writing a book as work, but I don't. In fact, I'm fortunate to say that I haven't worked since 1971, when I got my first paycheck as an automotive writer and photographer. Furthermore, I can attest that Peter Pan is alive and well, and he can be found pumping iron inside Ford flatheads and Chevy small blocks, and speed-shifting Hurst four-speeds and playfully turning roadster steering wheels, and knocking about town in tail-dragging cruisers and dago-axled highboys. To be sure, Peter Pan's spirit continues to burn because of people like you who refuse to relinquish the torch that was lit by the flame we know as hot rodding.

And it's that kind of childish spirit that makes my kind of work so much fun. Even so, I couldn't have played out this book without the

help of other people who, like me, love hot rods. Their contributions, big or small, made the completion of *Hot Rod Milestones* truly a fun and rewarding experience, to which I say, "Thank you, one and all."

By name, I'd like to thank: Ken Gross, Lesley Kendall, Mike Normile, Neal East, and the small cadre of volunteers and docents from the Petersen Automotive Museum; Greg Sharp, Steve Gibbs, and Wayne Phillips from the NHRA Motorsports Museum; Brian Brennan, Eric Geisert, Jim Rizzo, and the rest of the crew from *Street Rodder* magazine; Steve Alexander from *Hot Rod* magazine; Gray Baskerville from *Rod & Custom* magazine; Pat Ganhal, formerly of *Rod & Custom* magazine; Ralph Poole, from the old *Hop Up* magazine; Blackie Gejeian, Charles Gejeian, Chuck Krikorian, Dan Woods, Hank Becker, George Hood, Jim "Jake" Jacobs, Tony Thacker, Bruce Meyer, Pete Chapouris, Dick Kraft, Ron Roseberry, Rich Campos, John Athan, Tom Leonardo Sr., Tom Leonardo Jr., Ed Iskenderian, Art Himsl, Jack Quayle, Mike and Jo Sweeney, Ed and Karen Martin, Bob McCoy, D. Randy Riggs, Jim Losee and the kind people from Edelbrock Corp.; and finally ace photographer David Dewhurst (who showed me the light when it comes to photographing cars). If I overlooked anybody, I apologize, and you know who you are.

I cannot close without expressing my love and gratitude to my wife, Donna, and our two boys, Kyle and Chris. Their patience and help are surpassed only by the Big Hot Rodder who, in His own way, is a delight to have as my office partner.

I was fortunate to meet and interview a lot of interesting people while completing this book. Shown here are two hot rod legends in their own right: Blackie Gejeian (left), and Chuck Krikorian (driver's seat), sitting in the car that he built in 1959. Blackie owns *Emperor* today.

Introduction

"Classic hot rod." Now there's an interesting oxymoron because, for years, the words "hot rod" were synonymous with a particular single word—"jalopy." Indeed, based on what we hear, read, and see today concerning the origins of hot rodding, many of the early-era hot rods amounted to little more than jalopies that young men—boys in some instances—had built in their backyards and driveways, using discarded parts from discarded cars as their foundations. Eventually, by wit and by ways unknown, the throwaway pieces evolved into a car, rising again like the graceful phoenix from the ashes of ruin. In this case, a hot rod car was born.

Some of those early hot rods, it turned out, were better than others—judgmentally speaking, of course. As such, the mongrel beasts that eventually became known as hot rods began to assume

Want to see some hot rod milestones up close? One place to find them is at the NHRA Motorsports Museum at the Pomona Fairplex in Pomona, California. The museum houses dozens of great hot rods from the past, including street rods, dragsters, dry lakes and Bonneville racers, and oval track racers.

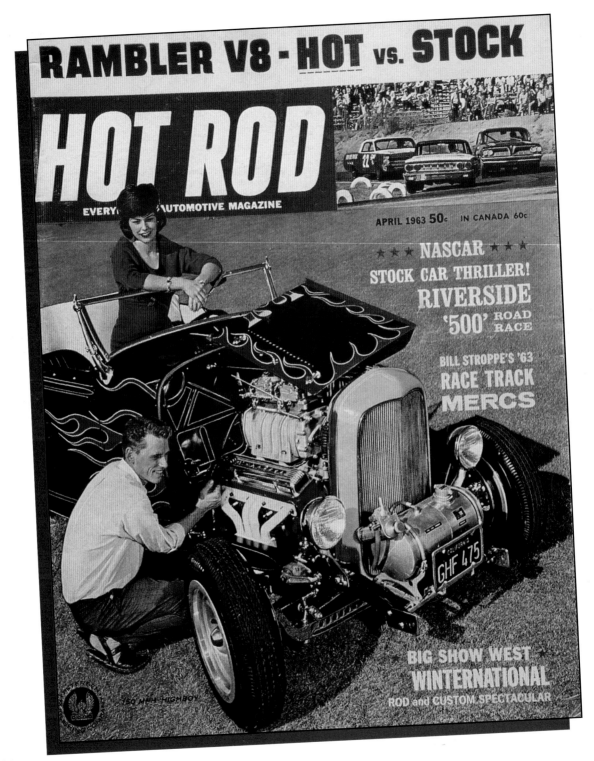

a pedigree of their own. Eventually, hot rodders and enthusiasts realized this, and soon recognized some of their "jalopies" as being truly unique, well-engineered cars. Hot rod engineering and construction techniques took a quantum leap forward after World War II when servicemen, having acquired training and experience relevant to hot rod building, returned home where they could once again indulge in their favorite pastime, that of building hot rods. And as the quality of cars improved, another trend evolved—promotion of weekend hot rod car shows where owners and builders (usually one and the same in those days) could showcase their talents.

Perhaps the hot rod show that had the greatest impact during those early years occurred in January 1948. In an effort to promote the sport of hot rodding and to turn the spotlight on the cars as truly bona fide, streetworthy vehicles, the SCTA (Southern California Timing Association) promoted a hot rod show at the Armory Building in Los Angeles, California. During that event another hot rod institution was born—*Hot Rod* magazine debuted at that Los Angeles show, and to this day the publication is regarded as the cornerstone for the sport and hobby.

Within two years hot rod car show fever had spread, and what was to become the largest and longest-running car show of its kind opened in Oakland at that city's Exposition Hall. The show eventually became known as the Grand National Roadster Show, and the inaugural event showcased 100 of the top hot rods, track roadsters, and custom cars in the country. The grand prize for the 1950 show's top roadster—in those days roadsters, above all other models, were typecast as hot rods—was a nine-foot trophy. The award became known as America's Most Beautiful Roadster (AMBR), and to this day it is presented to that show's best-judged roadster entry. The award is considered among the top honors in hot rodding. Several of the hot rods featured within the pages of this book are past AMBR recipients, and to be sure, an entire book could be devoted to the subject. Most recently I coauthored a book with Andy Southard Jr. that highlights the Grand National Roadster Show's first 50 years (*Oakland Roadster Show: 50 Years of Hot Rods and Customs*), and it was during research for that book that my editor at MBI Publishing, Keith Mathiowetz, suggested that I write a book focusing on some of hot rodding's milestone cars.

Given the topic's broad appeal, such an undertaking could be the equivalent of opening hot rodding's Pandora's box because, quite frankly, there are dozens, even hundreds of hot rods that qualify as milestone vehicles among the—may I use the term?—pedigree. Furthermore, all of hot rodding's milestone cars—in my slightly jaded, slightly objective opinion, anyway—qualify as classics. In this case, classic hot rods.

Due to the size and nature of this particular book, space and page count limit the number of cars to about a dozen. This, of course, means that some very high-profile and notable hot rods have been left out. Some candidates were omitted because they do not exist anymore; others were in deplorable condition during the production period of this book, so they could not be photographed; still others didn't make the cut because they were inaccessible to me. Finally, some hot rods aren't featured simply because I elected not to use them.

Every car featured in this book, with the exception of the Downey Brothers' 1926 Ford

What distinguishes an ordinary hot rod from a milestone hot rod? Several things, among them how the hot rod magazine industry recognizes the particular car. Tom McMullen's 1932 Ford roadster certainly warrants milestone status in several ways, including its appearance on the April 1963 *Hot Rod* magazine cover. McMullen built a clone of this car in the early 1970s, and *Street Rodder* magazine's staff was commissioned twice after that to build similar cars. The fourth edition of McMullen's famous roadster appears in chapter 12.

Model T, was in running condition when I completed the photography. The Downey car had been sitting for several years, though, only because its current owner, Hank Becker, ran the car so hard one day that the rare, one-of-a-kind single overhead camshaft Gemsa/Model T engine's innards finally gave way to the laws of physics. (Becker told me the old Model T—originally built as a hot rod in 1932—was going slightly faster than 100 miles per hour at the time the engine let go!) As a postscript, Becker vowed that he was going to get the classic hot rod back on the road before this book went to print. As he told me, "This (photo session) has inspired me to fix the engine and get her running again. Dain, I owe you." No, Hank, we owe you—once the car is back on the road!

One of the more memorable moments I experienced during the research for this book came when I photographed Chuck Krikorian's *Emperor*. Currently the car is owned by Blackie Gejeian, and for those of you who aren't familiar with Blackie, know this: he is a hot rodder first, a gentleman second. That's a compliment,

Famous hot rods of yesteryear are showing up in all parts of the country. This highly customized Model A roadster, known as *Little Candy Pearl*, was built by East Coast hot rodder Art Russell, who ultimately won the ISCA Grand National Championship with it during the 1965/1966 show season. After Russell's death the car was sold, and eventually fell into the hands of John Warunek who entered it at a 1997 Goodguys event, where *Little Candy Pearl* won Best Engine award. Later that year Ed and Karen Martin of Moscow Mills, Missouri, acquired it. The car remains in their possession today. *Ed and Karen Martin*

Just because a hot rod is considered an irreplaceable legend doesn't mean that it can't be driven on public roads! Art Bastian sits behind the wheel of the 1925 Ford originally built by Dick Kraft in the late 1940s. Kraft raced the car at the dry lakes until he sold it in 1954. The car resurfaced when Richard and Gary Seiden rebuilt it as the *Highland Plating Special*. The T roadster appeared on the March 1962 cover of *Hot Rod* magazine. Ron Weeks eventually bought the car, and later sold it to Mark Conforth, who passed it on to Bastian. *D. Randy Riggs*

because hot rodders are wonderful people. In any case, the first thing that Blackie the hot rodder did when I showed up at his raisin farm outside of Fresno, California, was to fire up the roadster's Cadillac engine for me to hear. The melodious tones that the six-carb, open-pipe engine made were music to my ears, and the hot rod kid inside me sang out in harmony with every rap of the throttle that Blackie gave the 38-year-old engine.

Every car in this book is an original, with the exception of the Tom McMullen roadster. As you'll read, this cool-looking hot rod is a clone. In fact, it's the fourth edition of the original that was built by McMullen in 1962. Tom built a similar Deuce highboy 10 years later as a project car for his new magazine, *Street Rodder*, and his third flamed roadster was a giveaway car that *SRM* promoted only a few years ago. The third clone—the car in this book—serves as

In the early days of hot rodding, few enthusiasts ever expected the cars that they built would become legends. A classic example is Vic Edelbrock Sr.'s 1932 Ford, shown here with Edelbrock and a crew member tending to the engine. For years Edelbrock used his Deuce highboy as a test bed for new products. Much of the testing was done on the dry lakes that dot the Southern California deserts. Later the car was owned by Ed Bosio, who rebuilt it and won the America's Most Beautiful Roadster award with it at the 1956 Oakland Roadster Show. The car returned to the Edelbrock stable and is currently under the watchful eye of Vic Edelbrock Jr. *Edelbrock collection*

the magazine's staff ride, and is driven by the magazine's editors at local events, as well as to and from work. (They take turns each week. . . . Can you believe it?!)

Some of the cars in this book have been restored or rebuilt, and others are, for the most part, original. That includes *Emperor*, which remains in the same livery as when Krikorian debuted it at the Oakland Roadster Show in 1960. The Buttera 1926 Ford Model T sedan appears to be original, too; however, it was rebuilt to original condition—but only after Lil' John Buttera and his wife, Joanne, used it as a daily driver for about eight years!

In the case of labeling the cars, I have taken the liberty of referring to each hot rod by its original owner. When possible I've traced its lineage through the years up to the current owner of record. But I feel that, above all else, credit must be given to the person or persons who built the car. In the case of the McMullen roadster, I've kept Tom McMullen as the original builder of record, even though the particular car featured in this book was built by the talented crew at Barry Lobeck's shop in Cleveland, Ohio. (For the record, Lobeck has built some remarkable milestone hot rods of his own.)

Also at this time I should point out that I have not included any modern hot rods, particularly of the "smoothie" or "high-tech" era. There are no Boyd Coddington originals, Don Thelan-built or recent AMBR winners. Instead I focused my attention on older cars because, for the most part, every car that is in this book has logged many miles on public highways. (That is not to say that some new hot rods aren't driven on a regular basis, because they are.) Ironically, the most recently built car (other than the McMullen clone by Lobeck) that I feature on these pages is a car that *looks* as if it were built back in 1947. I'm talking about Jim "Jake" Jacobs' *Jakeopage*, a 1928 Ford touring tub that was built in 1987 just for grins.

And it's Jake's nostalgia tub that brings us full circle to what hot rodding is really all about—building a driveable car out of what amount to throwaway parts. We, as hot rodders, have a choice: we can build high-dollar cars that look really neat and pretty—but, in some cases, so pretty and expensive that we are reluctant to drive them—or we can direct our energies to the creation of hot rods that we'll be anxious to drive on the road anytime, every time. I choose the latter. That's why I chose the cars that I did for this book. Whether or not you agree with my philosophy, I hope that you appreciate my selections.

Dain Gingerelli
Mission Viejo, California
February 1999

Flames and 1940s Fords go together like salt goes with pepper. And the most memorable flamed 1940 Ford of all time was Bob McCoy's Tudor that he built several decades ago. McCoy and his friends (Mike Root and Dick Fox), looking like a 1960s minstrel singing group, posed with the car during a photo session for *Hot Rod* magazine. Incidentally, a fellow by the name of Ray Cook applied the nosepiece flame job. *Bob McCoy*

Downey Brothers'
1926 Ford Roadster

An Early Soup Job That Helped Pave the Way for Today's Hot Rods

**Original builder: Downey Brothers
Originally built: 1932
Current owner: Hank Becker**

Hot rodders who submit to the adage, "They don't make them like they used to," will truly appreciate this 1926 Ford "soup job." The car was originally modified in 1932 by the Downey brothers, a couple of young hot rodders who lived in the Los Angeles area. Much of their original body and chassis work on this Model T remain intact, giving us a first-hand look at how hot rods were made in the days before the term "hot rod" had even been coined.

Originally built as a "soup job" in 1932, the Downey Brothers' 1926 Ford Model T is rolling proof that hot rodding's roots reach deep into time.

To be sure, hot rodding itself was in its infancy stages when the Downey brothers brought torch and hammer to their Model T so that they could personalize it to their liking. While there was no real accepted term to describe hot rod cars back in the early 1930s, many young enthusiasts referred to their modifieds as "soup jobs," a phrase most automotive historians feel was a derivative of "super job," which implied that the car was not an ordinary vehicle. Throughout time, the shortened phrase has been spelled supe job or soup job; we'll stick with the latter, as that seems to be the preferred spelling by many authors today.

The Downey brothers' car, now owned by Hank Becker of Garden Grove, California, was souped up using many of the techniques practiced during the early 1930s. The most noticeable treatment was to the chassis: the car was lowered six inches, and the wheelbase extended three inches. It's important to note too that, unlike today, there were no lowering kits or mail-order parts businesses from which to gather parts for these modifications, so the Downey boys did what any hot rodder did back in 1932—they took the car into the garage and performed the alterations themselves.

By raising the front and rear cross-members, the Model T was lowered several inches. The suicide front end extended the wheelbase a couple of inches, too.

The interior was recently—in relative terms—restitched with black leather. This Model T has been a hot rod since 1932, and was part of the Harrah's Auto Collection until sold at auction in about 1960.

The canvas spare tire cover was hand-painted in 1932, and remains original today. A curved, extended rear body panel adds dimension to the old hot rod's character.

In order to lower the car, they used traditional rodding techniques of the time. The rear portion of the body was dropped the desired amount by removing the third-member's cross-member, so that they could position a new cross-member higher in the frame. This also raised the axle higher into the frame, which had the effect of lowering the body. To make room for the higher cross-member, the Downeys notched a section of the trunk's floor so the center hump could poke through—a common practice. Ultimately, this procedure helped lower the body several inches.

The front was dropped by moving the front axle forward a few inches, then welding in a new cross-member to attach the front axle to the front of the frame rails rather than beneath them. Again, the desired effect was to lower the chassis and body. This type of modification came to be known as the "suicide front end" partially because some installations were so poorly welded that the entire assembly would disassociate itself from the car, causing an accident. Regardless of its structural integrity, the suicide front end on the Downey T lowered the flivver several inches. As a side note it's worth mentioning that the Downeys' welds have withstood the test of time.

Beyond these modifications the Downey boys wanted their soup job to sit even lower, so they determined that the 21-inch-diameter wheels had to go. In the interim they concluded that 18-inch wheels from a 1932 Plymouth shared the same lug bolt pattern as on their Model T, so they dedicated themselves to finding

A close-up of the suicide front cross-member gives a better understanding of the crude building techniques when compared to today's high-tech hardware.

a set for their soup job. According to lore the two young hot rodders turned to Midnight Auto for the wheels, where they picked up a semi-used set at the proverbial five-finger discount sale. In any case, the wheels, mixed with the new suspension, lowered the car six inches at both ends.

The suicide front end extended the wheelbase several inches, too, prompting the Downey boys to extend the running boards so that the fenders aligned with the altered wheelbase. The additional sheet metal was added to the front of the running boards, where pieces from another Model T were sectioned in and welded. The weld scars are still visible today.

To dress up the front fenders, skirts were welded to the rear portions of each fender, and the wheel openings were radiused so that each 18-inch wheel wouldn't appear abandoned inside a cavernous arch. Equal attention was given to the rear fenders, which were extended and flared to meet the protracted rear pan that was fashioned from sheet metal. These modifications were rudimentary by today's standards, but for their time they resulted in a very refined soup job.

Obviously the Downey brothers had an eye for aesthetics and proportions as evidenced by the chop job they gave to the Model T's windshield. To the untrained eye the tin lizzie's windshield appears to be of original dimensions, but in reality it has been shortened by sectioning a couple of inches from the upper and lower frames. "This," says Hank Becker, the car's current owner, "makes them look stock, but small and low."

The heart of any soup job was its engine. And so, to soup up their Model T, the Downeys installed a Rajo high-compression head that remained with the car for many years until Becker replaced the aging T motor with a Gemsa single overhead camshaft (SOHC) conversion. As Becker explains, "My good friend and mentor Joe Gemsa said I needed to run a single overhead cam of his design. He, with my help—although he didn't need my help!—built the engine that is in the car now, and is the only SOHC that Gemsa built for a Model T."

Another historical feature about the Downy T is its spare tire cover. According to Becker, the canvas cover was painted in 1932 by a Native American Indian whose name, through time, slipped into obscurity. Fortunately, the tire cover remained intact and with the car, and when the Model T joined the Harrah's Auto Collection in the late 1950s, the oil painting was preserved with the aging soup job.

Eventually Bill Honda acquired the roadster when the Harrah's collection was auctioned off following Bill Harrah's death. Honda primarily wanted the car so that he could use the canvas painting as a pattern to make the "Roar With Gilmore" collectibles that are sought among automobilia fans today. Once it served its purpose with Honda, the car passed hands again to Larve Thomas, who also owned a large collection of Cadillacs. George Hood, who worked for Thomas, took possession of the Model T in 1978, and it was then that the old flivver's body panels were massaged, and given its current coat of blue and black enamel paint. Hood owned the Model T for a couple of years, eventually trading it to Becker for another old Ford hot rod.

Shortly after Becker acquired rights to the old soup job, he and Gemsa installed the one-off SOHC Model T engine. The car remained a part of the Southern California hot rod scene for a number of years until Becker parked it so that he could devote time to other projects. When Becker took the car out of mothballs for these photos, he announced that the car would be prepped and ready to be back on the road where it belongs. "After all," exclaimed Becker, a die-hard hot rod enthusiast who enjoys the old stuff, "this car is too important to keep in my garage—or in a museum!"

Originally the Downey brothers equipped their Model T engine with a Rajo cylinder head. New owner Hank Becker helped Joe Gemsa build the single-overhead camshaft conversion that powers the car today. This is the only SOHC that Gemsa made for Model Ts.

Chapter 2

John Athan's 1929 Ford Roadster

One of the First A-V8 Roadsters Ever Built

Original builder: John Athan
Originally built: 1937
Current owner: John Athan

Perhaps one of hot rodding's more colorful A-V8 roadsters is John Athan's 1929 Ford, best known as *The Elvis Car* because it appeared in the movie *Loving You*, starring Elvis Presley. John drove the Model A for nearly 40 years until he retired the roadster to a shed on his property in Culver City, California. And there the A-V8 sat for nearly two decades,

John Athan's 1929 Ford Model A roadster is best known as *The Elvis Car* because it co-starred with the King in the movie *Loving You*. This was among the first Model A roadster hot rods to have a flathead V-8 engine and Deuce rails.

until it was hauled out for restoration by Tom Leonardo Jr. in 1996.

The prolonged convalescence had not been kind to the old Ford. The shed's leaky roof dripped rainwater on the roadster's leather upholstery, and careless co-tenants—rodents and spiders and other creepy creatures—assaulted it from below. When Tom retrieved the roadster from its slumber, the brakes were seized and the deflated tires stuck steadfastly to the floor.

"We had to drag it out of the garage with a chain and a rope," said Tom. In the process the left front tire's valve stem ripped off; as a quick fix Tom jabbed a ballpoint pen into the stem hole so the tire would hold air until they got back to the shop.

Actually, the pen-patch incident is just one of many interesting episodes involving this hot rod. The A-V8's saga began in 1937 when John built the car. He paid $7.00 for the body and $5.50 for the Deuce frame. To the best of John's knowledge, this was among the first Model A bodies set on a Deuce frame.

John was 19 years old at the time and he was running on high-octane hormones, doing many of the irresponsible things that footloose young men seem to do. It didn't help that he kept company with another young rapscallion, a kid named Ed Iskenderian. They had been high school classmates, and together jumped feet first into hot rodding shortly before graduation. Moreover, they have remained friends ever since, and they meet every day with several other notable yesteryear hot rodders—Nick Arias, Louie Senter, Kong Jackson, and Harold Johansen, among them—for lunch and coffee. That's when they'll

John Athan completed the car in late 1936. The A-V8 served as his daily driver for many years and raced several times at the dry lakes. In fact, an original El Mirage Dry Lakes timing slip recently was retrieved from the trunk! Date: May 3, 1942. Speed: 108.50 miles per hour.

Bon-A-Rus restored the interior with red vinyl. The Auburn dash with engine-turned gauge insert was cleaned up, but otherwise remains the same as when Elvis Presley stared down at them 40 years ago.

trade tall tales and true lies about the old days, too. As you might suspect, some of the lore passed around the lunch table during the afternoon bull sessions involve John and his old A-V8.

One of John's favorite tales is about the time that he took a perky teenage girl for a ride through Beverly Hills. Out to impress the shapely young coed, John accelerated hard from an intersection. The sound of spinning tires also caught the attention of a nearby police officer who gave chase. Rather than pull over and face the consequences, young John raced through the city, and when he finally got clear, he sought shelter in a vacant garage alongside a neighborhood house. The people inside the house either didn't hear him drive in, or were too frightened to do anything about it. In any case, John slammed the garage door shut, then waited inside with the young gal—now a fugitive too—until he was sure the coast was clear.

"Finally," John said, telling his story today, "I figured the people inside (the house) might come out to see what the commotion was all about. So I told her (the girl) to open the door and jump in after I started the engine. We did, and I backed out as quickly as I could, but there were no cops around. But I tell ya, I kept a low profile for a long time after that!"

That wasn't the only incident involving young John, his 1929 roadster, and a pretty girl. Once, while showing off to another young lass on board, John jabbed the throttle pedal down too quickly exiting a corner, forcing the car into a complete 360-degree spin. Miraculously he saved it from spinning again, and continued to motor down the road as if the little stunt were preordained. "I didn't do it on purpose," recalls John, "but when we got back to the drive-in (restaurant) she told them (their friends) what happened. 'Boy,' she told everybody when we got back, 'is he a good driver.' I didn't tell them that I didn't do it on purpose."

Most of John's memories with *The Elvis Car*, though, include his best pal, Iskenderian. Back in those days, putting John and Isky together was like mixing nitro with glycerine. The effects were always unpredictable, and John and Isky usually managed to do something . . . interesting. Such as the time John and his A-V8 got pulled over at the corner of La Brea and Melrose by the L.A. police because the roadster's headlights were too low. The law said 32 inches from

The baloney-cut aluminum intake stacks were made by Athan about the time the car was first built. Frank Oddo photographed the car for *Street Rodder* magazine in 1977.

the ground. Anything less and it was considered dangerous. The lights on John's car were slightly more than 29 inches above the pavement. Busted. "They yanked me out of the car, and had me stand up against the wall with my hands high. They wanted me to look like a big criminal in front of those people (passersby)." In any case, the citation turned out to be a fix-it ticket, which meant that all John had to do was correct the mechanical problem, then show it to an officer of the law who would sign off.

Enter Ed Iskenderian. He suggested, as only a true best friend can do, that rather than reposition the headlights—Athan was going to put them back to their cool-looking 29-inch height once he got signed off anyway—they temporarily elevate the car to make it *appear* as though the problem was solved. So Isky slipped a pair of 2x4 sections of lumber under the frame rails, raising the headlights another couple of inches off the ground. As they approached the police station for inspection one of the pieces of wood slipped out, causing the car to sag on one side. As John told this story, "Remember, this was Ed's idea. Anyway, the policeman asked what the problem was, so I told him that we were having a little trouble with the motor mounts." The cop bought the bogus explanation, and dutifully signed off the ticket.

Then there was the time John and Isky were driving to Muroc Dry Lake for a speed trials event. John had installed a new dropped axle onto the car a few days before, and while traveling about 80 miles per hour he hit a big dip in the road. The force snapped the welded axle, and when it broke the car immediately embarked in a series of 360-degree spins on the dirt shoulder. Finally the car and its two befuddled occupants came to a wild, yet safe stop. Said John, "I didn't think the dust was ever going to stop settling on us!" (As a side note, during the restoration Leonardo unearthed a pair of timing slips in the car that documented a weekend of racing at Muroc Dry Lake on May 3, 1942 by John and his roadster. John drove the car 104.77 miles per hour on the first pass, and 108.50 miles per hour on the second run. The event was sponsored by the Road Rebels, and John's entry was number 225.)

John continued to drive his A-V8 roadster after World War II, and into the 1950s when

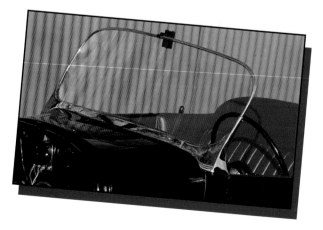

Perhaps the most notable feature on *The Elvis Car* is the curved windshield. Athan formed it from a 1940 Chrysler rear window. The bottom of the frame was completed in 1998 when Tom Leonardo Sr. restored the car.

hot rodding became popular across the country. It was about then that his highboy earned its 15 minutes of fame when Paramount Studios selected it to be the car that Deke Rivers—played by Elvis Presley—would drive in the 1957 movie *Loving You*.

The car appeared in several other movies, among them *I Want You and A Woman Of Distinction*. John and the roadster also signed up for a short documentary about gas stations. Hot rods, it seems, have always been good for sprucing up flicks.

John's car resurfaced in 1978 for a photo feature in *Street Rodder* magazine's March 1978 issue. Frank Oddo photographed the car, and it wore the same chrome-spoke wheels and redline tires in those photos as when it embarked on its extended slumber in John's leaky garage shed.

Which brings us to the reincarnation of *The Elvis Car*. "I had been bugging him and bugging him for 10 years to restore the car," said Leonardo. "Finally he said, 'okay,' and we got busy on it."

Obviously John and Tom didn't want to change the character of the car. Tom also convinced John that the roadster should be built as near as possible to the condition it appeared in *Loving You*. That meant mounting 16-inch Kelsey-Hayes in place of the chrome-spoke wheels. The wide whitewall Firestones are 5.60-16 front and 6.50-16 rear.

For the most part the restoration simply required replacing old with new. For instance, the body was scraped to bare metal (in the process revealing the patchwork on the driver door, the result of a dent that the King himself caused during a chase scene in the movie; Presley did most of his own driving in *Loving You*), then repainted black. Tom tried to enlist Bob Po, who originally upholstered the car back in the 1930s, to re-skin the interior. But Po, now retired, declined, so Tom had his favorite trimmer, Tito at Bon-A-Rus (Orange, California), restore the interior with fresh red vinyl.

Tom also took the time to finish what John had started so many years ago—he completed

Ever wonder what an old hot rod looks like during its restoration? Here Leonardo Sr. (left) points out a few body repairs to owner Athan (center) and his good friend Ed Iskenderian. As you can see, the body was stripped to bare metal, revealing a repair patch to the lower left door due to a driving accident that Elvis had during filming for *Loving You*.

the windshield frame. Look closely at early pictures of *The Elvis Car*, and you'll see that the bottom of the frame holding the glass (originally a 1939 Chrysler rear window) is missing. That's because John never finished it. To do the job, Tom took pieces of scrap metal from his inventory, then welded and pieced together a section to fit. It wasn't an easy task, but Tom felt that it was time this unique curved fixture be completed.

Fortunately the flathead V-8 didn't require a rebuild, although the upper chromed radiator tubes were replaced, and the carbs cleaned and tuned. There was never consideration to replace the stock F-heads with high-compression aluminum heads, either. The goal was to match the car to its movie appearance, so the block and heads were merely painted red.

A week after it was finished, John's highboy rolled into the National Hot Rod Association (NHRA) Motorsports Museum in Pomona, California, for that facility's opening night. But only after *The Elvis Car* posed in front of my Nikon one more time. After all, this car is a star, and its fans deserve to see it one more time.

Ed Iskenderian's
1924 Ford Roadster

A Hot Rod for All Time

Original builder: Ed Iskenderian
Originally built: 1940
Current owner: Ed Iskenderian

It's rusty, it's dusty, it's even a little bit crusty, but the fact remains: Ed Iskenderian's 1924 Ford is among the early paragons that influenced hot rod builders in subsequent years. Originally built in 1940, Isky's roadster, or *Cucaracha* as it was known among the 1950s hot rod community, has "been there, done that, and been around the block again." This Model T roadster served as Ed's daily driver throughout the 1940s; was timed on a Southern

The Isky roadster's design, emphasized by the long sidepipes along the Essex frame rails, is as unique today as it was in 1940. Other mongrelized parts include Plymouth hydraulic brakes, Franklin steering, and Pontiac grille halves.

California dry lake bed in 1942 at 120 miles per hour; and it appeared as *Hot Rod* magazine's sixth-ever cover car. If that's not enough, 25 years later the editors of *Hot Rod* paid homage to it one more time to help celebrate *HRM*'s silver anniversary, and *Rod & Custom* featured it 20 years after that with a full photo feature highlighting the car's design intricacies. Finally, it was invited to be among the first exhibits at the NHRA Motorsports Museum in Pomona, California.

Let's turn the clock back to 1948, so that you can gain a better appreciation for Isky's roadster from *Hot Rod*'s original write-up in the June issue. In conclusion, the magazine article stated: "The June Hot Rod of the Month rates admiring comment wherever it goes. But, believe it or not, Ed bought the whole '24 T from which the body was made for only $4.00. That was in 1939. Just try to buy it now."

As things developed, Ed never sold the car. With the success of his camshaft business growing every day during the 1950s, the Model T soon found itself in the back of the shop where it collected dust until *Hot Rod* magazine's Steve Alexander ventured into the nether reaches of Isky's shop to report on the old hot rod's status for *HRM*'s silver anniversary issue. For the January 1973 issue, Alexander wrote: ". . . We passed the 'RESTRICTED AREA: AUTHORIZED PERSONNEL ONLY' sign and into a giant room full

Ed Iskenderian's roadster appeared on the June 1948 cover of *Hot Rod* magazine. In 1998 the car took part in the NHRA Motorsports Museum's grand opening.

Like a favorite easy chair, the interior of Isky's roadster is worn with age. Note the column-mounted tachometer beneath the banjo steering wheel. Truly a classic interior!

This is the view that Ed Iskenderian had when he and his V-8-powered roadster rumbled through the Western Timing Association's speed traps at 120 miles per hour across El Mirage in 1942.

of engines, motorcycles, drill presses, engine stands, heads and manifolds, a sandblaster, boxes full of pistons and parts, a lawnmower, a soft-drink machine, an old record player and more 'stuff' piled high . . . The roadster sat there with its paint dull and chrome pitted."

Twenty years later the roadster surfaced again, the first time at the Los Angeles Roadsters annual show in Pomona, and later at the first NHRA California Hot Rod Reunion at Bakersfield Raceway. Both times Isky, being the casual guy that he is, didn't bother to rub off the dust, or try to remove much of the rust from the old car. He simply drove up to the gate at Pomona, seeking entry into the most famous roadster show in the country. And he was promptly refused admittance because the car looked . . . neglected. Which it was. And the L.A. Roadster Show has a reputation for neglecting neglected hot rods.

The incident caused a minor uproar among purists, and many of the old-timer L.A. Roadsters members were somewhat miffed and embarrassed by what happened. The matter eventually was reconciled, but most of all it helped rodders of all generations to gain an appreciation of, and to foster insight for, some of the more historic cars that survive and, more important, remain intact today.

The second coming, this time at Bakersfield Raceway, was witnessed by Pat Ganahl, then editor for *Rod & Custom* magazine. Sighting Isky's roadster at Bakersfield that day inspired Ganahl to arrange a photo feature of the car in a forthcoming issue of *R&C*. The article appeared in the April 1993 issue, and Ganahl wrote: "When I saw Isky come tooling around the corner of the grandstands in his black T roadster at the Bakersfield Reunion, I just about dropped my camera on the track. I remembered a couple of times—once in the mid-'50s and again about 20 to 25 years ago—that some magazine guys talked him into getting it out, cleaning it up, and taking it for a spin for a story. I had seen it several times over the past 20 years, sitting in a storage area behind the dyno room at Iskenderian Cams, covered with cardboard, collecting

The signature to the Isky roadster is the engine, a Ford flathead V-8 with Maxi heads. These heads were originally designed for commercial trucks in order to minimize engine overheating. The F-head design uses the stock intake valve with an overhead exhaust valve. Iskenderian inscribed the valve covers himself.

Iskenderian's pride and joy? The skull-and-wings radiator ornament that he cast as a high school student in an industrial arts class at Long Beach (California) Poly High. That would have been shortly before World War II.

dust, and rusting." And so the story of Isky's roadster continues:

When he built the car in 1940, Isky based his hot rod on a pair of Essex frame rails. Essex frames were plentiful during the prewar years, and the stout fixtures were popular among racers because the rails were strong. In the case of hot rodders, the beefy Essex rails also made it convenient for strapping in a Ford flathead V-8 engine, a motor that was growing in popularity among hot rodders by the time Isky built his T roadster. The car was obviously well built, and has steadfastly withstood the test of time, time and time again.

Doane Spencer's
1932 Ford Roadster

The Quintessential Highboy Roadster

Original builder: Doane Spencer
Originally built: 1946
Current owner: Bruce Meyer

Never has there been a hot rod that garnered more accolades and praise from enthusiasts, yet never won a major car show or award, like the Doane Spencer 1932 Ford roadster. Fact: The Spencer highboy was never a candidate for the America's Most Beautiful Roadster award presented every year to the top gun at

The words "elegance" and "hot rod" weren't meant to complement each other—except when you are talking about the Doane Spencer hot rod. In its restored condition, the Deuce highboy became the first recipient of the Pebble Beach Concours d'Elegance, Hot Rod Division, 1997.

When hot rodders speak of "the look," they are referring to the car's stance and the attitude that prevails. The Doane Spencer 1932 highboy has "the look."

Perhaps the popularity of the classic DuVall-style windshield among hot rodders today can be traced to the Doane Spencer highboy. Spencer mounted his original DuVall windshield to the Deuce's cowl in 1946.

To prepare the car for the LaCarrera-Panamericana Road Race in Mexico, Spencer mounted a Gordon Schroeder steering box. This set the drag link parallel to the ground to help minimize bump steer.

the Grand National Roadster Show, and for that matter this famous Deuce roadster never won a sweepstakes at any major car show during the 1950s. Nor was the Spencer highboy a class record holder at the dry lakes speed trials, and it didn't set the world on fire at the quarter-mile drag strips, either.

Despite its modest trophy collection, this highboy remained an icon of hot rod lore, treated with revered awe by hot rodders everywhere. Part of the car's charisma might have had something to do with Spencer's rather brash, even arrogant, personality. See, Spencer, who died of cancer in 1995, was a no-nonsense person, one who typified the American pioneering spirit. He stuck by his convictions and, right or wrong, he spoke his mind based on his beliefs. Mike Normile, a young hot rod builder today who worked several years for Spencer until his death, described his mentor as a man with unmatched self-confidence. "When it came to hot rods, Doane would tell it like it was," said Normile, "and he was a technical genius."

And it was Spencer's technical genius, matched with an uncanny creative brilliance, that led to perhaps the most talked-about Deuce highboy roadster of all time. The car's origins as a fenderless highboy can be traced back to 1937 when a hot rodder by the name of Jack Cort began the project. Not until after World War II, when Spencer acquired the unfinished highboy, did it become the classic hot rod as we know it today. Once Spencer put the highboy on the road, the fenderless roadster established itself as one of Southern California's premier hot rods. Spencer, a member of the Glendale Stokers, displayed the car at several of the early SCTA-sponsored car shows in the Los Angeles Armory, and he often could be found at the nearby dry lakes where the flathead V-8-powered highboy posted respectable, if not stellar, times.

By 1953 Spencer had become enraptured by the annual LaCarrera-Panamericana Road Race, first held in Mexico in 1950. Smitten by tales of south-of-the-border wide-open racing from such local go-fast crazies as Ray Brock and Ak Miller, Spencer got hooked on the idea of converting his highboy hot rod into a highboy road racer. So he disassembled the car to install a new overhead valve Y-block Lincoln V-8 engine for more power, and he re-engineered practically the entire chassis to improve handling for the race. The makeover included a Gordon Schroeder steering system to help reduce bump-steer, rebuilding the solid-axle suspension, and routing the exhaust system to exit through the sides of the tall Deuce frame rails so that he could gain additional ground clearance over the treacherous Mexican roads. Unfortunately—or fortunately, if you listen to some hot rod aficionados today—Spencer didn't complete the transformation in time for the 1955 race, the final year of the event (spectator safety led to the demise of the 1,000-mile race over public roads).

Disenchanted with the ordeal, Spencer's interests changed slightly at that time, too. He was married, and he wanted to buy a house. To finance the venture he sold many of the Deuce highboy's speed parts, and about a year later—in 1956—he sold the body, frame, and Valley Custom hardtop to *Rod & Custom* magazine editor, Lynn Wineland. Unfortunately (or fortunately, again) Wineland wasn't able to complete the project, and the fabled roadster sat unfinished in

Pete Chapouris' SO-CAL Speed Shop restored the car in 1997. Even though the chrome-plating is new, the brake backing plate's air scoop and exit holes are what Spencer engineered five decades ago.

his garage until 1969 when another *R&C* scribe, associate editor Neal East, assumed ownership.

It was East who got the car back on the road, in the process slipping another flathead V-8 engine between the frame rails. Back on the road again, the ex-Spencer car could be found with East behind its classic racing roadster steering wheel at all sorts of hot rod events. East raced it at the vintage road races at nearby Willow Springs Raceway, and in 1976 he drove it back east to the National Street Rod Association (NSRA) Street Rod Nationals, held at Tulsa, Oklahoma. "But I never had time to truly finish the car," recalled East, "so I took it apart in 1985 to do just that."

Unfortunately (or fortunately, again!) East never saw the restoration to its completion, and 10 years later hot rod collector and diehard enthusiast Bruce Meyer purchased—through Spencer's encouragement, it's worth adding—the old warhorse. Meyer wasted little time shucking the car over to Pete Chapouris (formerly Pete Chapouris Group, now called SO-CAL Speed Shop) in Pomona, California, for restoration.

The Chapouris Group completed the car in time for the 1997 Pebble Beach Concours d'Elegance, where the car was shown. Through encouragement from members of hot rodding's tradition-minded establishment, the promoters at Pebble Beach had agreed to include American-made hot rods in their own class for a trial year. Typically a show for purebred pedigree cars, the stately show grounds on the Pebble Beach golf fairway included, for the first time ever, the hot rod mongrels—cars bred from undocumented origins but nevertheless classic and timeless in their design and construction.

Only a handful of hot rods were invited for the debut at Pebble Beach, among them the Doane Spencer roadster, entered by Bruce Meyer. When the judges had tallied up their scorecards, the hot rod with the highest marks was the Deuce highboy that, until that time, had never won a major award in hot rodding! The

Tom Sewell trimmed the interior with leather. Again, the word "elegance" enters the equation. The interior's simplistic beauty mirrors that of the entire car, making this truly a landmark hot rod of all ages.

Doane Spencer car had finally achieved the recognition that it truly deserved.

When the Petersen Automotive Museum opened its wing for hot rods, Meyer stepped forward one more time to sponsor the dedicated floor space. Fittingly called the Bruce Meyer Gallery, in 1997 the Spencer car became the centerpiece for the hot rod display.

Today the Doane Spencer car is the showcase at rod runs and car shows across the country. Meyer drives the car occasionally, too, and if the car has no commitments it returns to its place of honor at the Petersen Museum.

Do hot rods need to be given their accorded due if they are to be considered milestones of the hobby? Not really, but in the case of the Doane Spencer highboy, the presentation as top dog at the Pebble Beach fandango was long overdue. Doane Spencer would have been proud, although if pressed for comment he probably would have curtly replied, "It's about time."

Chapter 5

Bill NeiKamp's
1929 Ford Roadster

Like No Roadster Ever Before—or Since

Original builder: Bill NeiKamp
Originally built: 1949
Current owner: Jim "Jake" Jacobs

The difference between first and second place is more than just a matter of deciphering who won and who lost. There's an element of name recognition involved, too. And, generally speaking, more recognition goes to the guy who's first, not second. Indeed, history doesn't record who the second-ever European to see America happened to be, nor is the name Aldrin (the man who followed

One thing that can never be taken away from the Bill NeiKamp roadster is its honor of being the first to win the America's Most Beautiful Roadster award. Jim "Jake" Jacobs restored the car to its former glory, and today the car sports the same livery as when it won the AMBR in 1950.

in Neil Armstrong's lunar footsteps; and can you even recall Aldrin's *first* name?) exactly a household word. Ditto for the second man to break the sound barrier after Chuck Yeager smashed through in 1947, or Sgt. Joe Friday's dutiful sidekick on the 1950s television show *Dragnet*.

The same recognition honor code applies to many of hot rodding's first-place finishers, too. In particular, when the topic at a bench racing session drifts to any of the 50 past winners of the America's Most Beautiful Roadster (AMBR) award, presented annually at the Grand National Roadster Show, any hot rodder worth his lug nuts knows that the *first* winner happened to be a 1929 Ford roadster belonging to a 44-year-old auto body repairman named Bill NeiKamp.

In hot rod lore that car has come to be known as the *NeiKamp Roadster*, even though the car was never officially christened as such.

For the record, NeiKamp (pronounced *knee-camp*) built his A-V8 in the garage. He constructed the award-winning roadster by hand, and to maintain the structural integrity of its steel, any cutting that was necessary was done with a hand-held hacksaw. As NeiKamp told current owner Jim Jacobs in 1971 for an article Jake penned for *Rod & Custom* magazine that documented the car's first restoration, "torch cutting and arc welding speeds up oxidation in the metals, so I patiently cut everything by handsaw."

Actually, the car adhered to many of the usual styling and building trends that rodders

Bill NeiKamp built the entire car himself. He formed the rear roll pan and skirts shrouding the Essex frame rails, he fashioned the nerf bars, and he had final say on the placement of the dual exhaust pipes.

followed back in 1949. For instance, the channeled body sat atop 1927 Essex frame rails, which happened to be convenient and plentiful (the Essex frame was strong, and its rails had a rear kick-up much like that of a 1932 Ford frame, making it suitable for channeled Model A's, such as the *NeiKamp Roadster*), and a Model A rear cross-member formed the terminus for an early all-Ford drivetrain. A set of hydraulic brakes from a 1940 Ford was pressed into service, and NeiKamp hand-formed the bellypan, floorboard, bullet nose, and three-piece hood himself.

Under the hood rested a 1942 Ford V-8, although that was eventually replaced by the car's third owner, who decided in 1958 that what the car really needed was a newer nailhead overhead valve Buick V-8. More on that later. Back to 1950, when the '29 garnered first bragging rights to the nine-foot-tall AMBR trophy at Oakland:

In a report of the 1950 National Roadster Show penned for *Hot Rod* magazine, Griff Borgeson wrote about NeiKamp's roadster: ". . . (the car is) neither flashy, spectacular or even tested for speed. What *is* remarkable about the car, and the feature that enabled it to take seven other major awards, is the perfect purity of its layout and workmanship. Everything about the job is subdued and—there's that word again—perfect."

Later that year NeiKamp entered the car in the Pasadena Reliability Run (PRR) where it was honored again for its workmanship. It shared the

Nothing trick about the front suspension—a 1937 Ford tube axle, 1940 Ford spindles, and juicers. Craftsmanship, though, was superb, including the hand-formed nose and grille, and the sprint car nerf bar.

PRR's Best Appearing award in 1951 with another hot rod, and before the 1950 calendar was taken off the shop wall NeiKamp had raced his roadster at the El Mirage Dry Lake speed trials. Eventually the car posted a top speed of 142.40 miles per hour (July 13, 1952). This was the roadster's final attempt ever on the dry lakes, and today Jake points out that he has all 12 timing tags that the car earned during the three seasons (and four events) that NeiKamp campaigned at El Mirage. Beams a proud Jake today: "Most guys are excited to have one old timing tag for their car. I've got every one for this car."

In fact, Jake has all the trophies and dash plaques the car ever won, and NeiKamp even passed along his original receipts and other

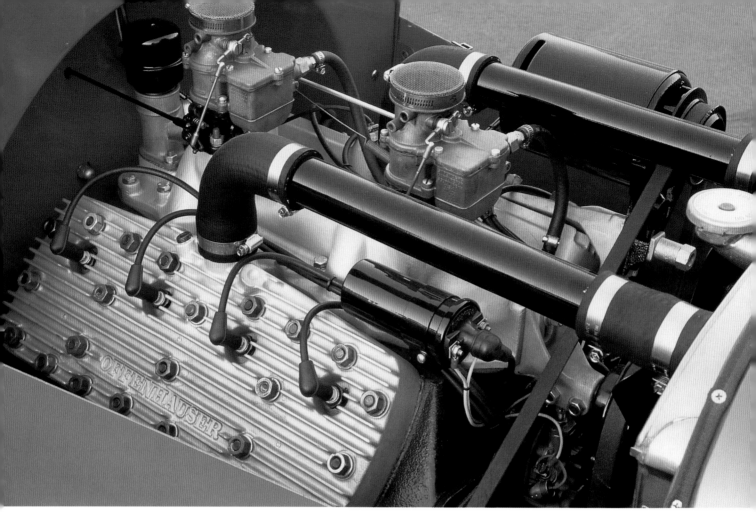

The original 1942 Mercury V-8 had Evans cylinder heads, Weiand intake manifold, Winfield cam, and Kurten ignition. That motor was eventually replaced by a 1957 Buick V-8, another flathead, and later a Chevy small-block. Not until 1997 when Jake showed the car at the Pebble Beach Concours d'Elegance was the car given this flattie V-8.

documentation that today help chronicle the roadster's genesis as a hot rod. As a side note, Jake can tell you that NeiKamp paid $15 for the Model A body, or that the entire project set its original owner back $1,888.72.

A couple of years after it won Oakland, someone offered NeiKamp $2,800 for the roadster, but he turned it down. Instead, NeiKamp raffled the car off in 1952 to raise funds for a friend who had been seriously injured in a crash at Bonneville that summer. Raffle tickets sold for $1.50 at the Santa Ana Drags (ironically raising only $700), and the holder of the winning number was a young soldier home on leave named Dick Russell.

After his discharge from the military, Russell drove the car for several years. The roadster served as his daily driver, and he raced it a few times at the Santa Ana Drags. Finally he sold it to Dylmer Brink who was responsible for plucking the old flathead V-8 motor out and inserting the Buick. Fortunately, as far as hot rod historians are concerned, Brink never completed the engine swap, and the car sat in disrepair until 1969 when Jake bought it.

Jake was associate editor for *Rod & Custom* magazine at the time, and he immediately recognized the forgotten old car as the *NeiKamp Roadster*. Having located the owner, Jake opened negotiations to buy the former AMBR winner. "He turned down my original offer," Jake says today, "and told me that he'd give it to his son before he sold it for the price I offered. So I had to step up and put in the extra $300." The total selling price: $1,300.

Should you harbor nasty notions that Jake all but stole the car, hear him when he says: "But that was at a time when nobody cared about these kinds of cars (old hot rods)." He tactfully added, "We were putting Jag rear ends under stock-looking cars, that sort of thing."

Jake restored the car in 1971. He equipped it with another Ford flathead V-8, but that engine expired when he was driving home from—of all places—the Oakland Roadster Show where the '29 had been on display to help commemorate the show's 25th anniversary. As a quick fix he installed a small-block Chevy, which remained part of the roadster's drivetrain until 1997, when Jake entered the now-historic car in the Pebble Beach Concours d'Elegance.

The 1997 Pebble Beach show was the first time that hot rod mongrels had been included in this otherwise pedigree affair, so Jake groomed the car to its original 1950-AMBR appearance. Back into the engine bay went a flathead V-8. In fact, Jake decided that the car should appear as close to its 1950 show-winning form as possible, so off came the windshield, and on went the tonneau cover. (Jake points out that most people today are more familiar with the *NeiKamp Roadster* in its 1951 livery, with windshield intact and no passenger-side tarp.) In his words about preparing for Pebble Beach, "I backdated it more for that event," although the car's interior still needed its own retrofit to 1950 specs.

Regardless of how pure the car is or isn't, the fact remains that the *NeiKamp Roadster* has survived more than half a century as a true hot rod icon. That, above all else, gives hot rod purists and elitists the latitude to revel in the *NeiKamp Roadster*'s existence. After all, it was the first car to win the most prestigious car show award ever conceived, which makes this roadster more special than at least 49 other hot rods ever built.

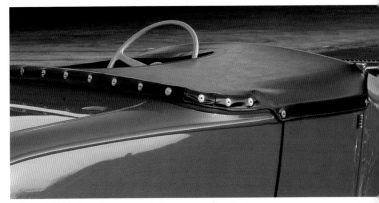

When Bill NeiKamp entered the car at the first National Roadster Show in 1950, he displayed the car as a dry lakes racer—windshield off, tonneau cover on. That is how Jim Jacobs shows the car today, in its 1950 AMBR-winning form.

The dashboard is slightly different than the original; NeiKamp used a three-gauge instrument cluster from a 1949 Plymouth, while Jake formed his own dash using Stewart-Warner gauges. NeiKamp also used a 1941 Ford steering wheel, while Jake's car has a classic 1940 Ford wheel.

Chapter 6

Dick Kraft's
1924 Ford Model T

The Evolution of Today's Rail Dragsters Began 50 Years Ago with *The Bug*

Original builder: Dick Kraft
Originally built: 1950
Current owner: Dick Kraft

It took a free spirit the likes of a certain young hot rodder named Dick Kraft to build *The Bug*, for this was a car that distilled the very essence of hot rodding into its most basic form. *The Bug* was built in 1950, and is among the first hot rods built expressly for drag

The Bug was simple in its creation: a souped-up Ford flathead V-8 was slung from a Model T frame; the gas tank (from an insecticide spray can, thus the name *The Bug*) was mounted next to a single seat for the driver; and a cowl section from a 1926 Model T held the steering column in position.

racing. Indeed, half a century later purists herald *The Bug* as the predecessor to the rail dragsters that ply the quarter-mile strips throughout the country today.

To appreciate *The Bug*'s significance in history, though, it's worth noting that in 1950, when this modified Model T first rolled onto the race track, hot rodders had been merely dabbling in drag racing. Up to then most of the racing was confined to impromptu, and highly illegal, showdowns on backroads and deserted streets, with occasional one-day organized meets held on abandoned airstrips and parking lots. As you can imagine, in 1950 there was no clear definition about what drag racing really was all about; there were no national organizations such as the NHRA to offer guidance for promoters and racers to follow, nor were there any dedicated drag strips where young hot rodders could go to test the acceleration of their cars. Not until the weekend of July 2, 1950, when the Santa Ana Drags opened on a

"Necessity is the mother of invention," so the saying goes. And through the necessity of accelerating quicker in order to beat a certain motorcycle dubbed *The Beast*, a young racer named Dick Kraft built what many consider today to be the first rail dragster, known back in 1950 as *The Bug*.

Reasoning that the engine wouldn't overheat before the quarter-mile stretch of race track ran out, Kraft decided to forego the radiator. To seal the cooling system he plugged the end with plumber's pipe. The mother of invention won again.

weekly basis, could drag racing qualify itself as an organized form of motorsport.

The establishment of a weekly program marked the beginning of a new era for hot rodding, too, because now Southern California rodders didn't have to make the long trek to the desert dry lakes to race their "soup jobs." Instead, they could drive a few miles to the vacated west runway of nearby Orange County Airport on the outskirts of Santa Ana, California, where every Sunday morning they could pay the one dollar entry fee to compete in acceleration contests. Best of all, the Santa Ana Drags were rather casual, and there was no class structure or complex safety tech inspections to bog down the program. Quite simply, once an informal tech crew examined your car's tires and brakes, you could position your car in the staging lanes to await your turn for a pass down the strip (early Santa Ana Drags included rolling starts, and racing was confined to one-third-mile passes). At the end of the day, the two fastest competitors squared off for Top Eliminator honors. It was that simple.

Among the first to challenge for Top Eliminator was Dick Kraft and his T-V8. At the time his fenderless Model T was like many others that could be found competing on the dry lakes—the souped-up flivver was stripped of all superfluous bodyware save the T-bucket itself, and the spindly

A frontal view illustrates how simple and direct Kraft's design was. The day that he rolled this car onto the drag strip at Santa Ana Airport, another branch sprouted on the hot rod family tree.

contraption was powered by a Ford flathead V-8. The car was fast, and initially the only thing that prevented Kraft and his T-V8 from scoring Top Eliminator honors the first couple of weeks at Santa Ana was a motorcycle ridden by Al Keys.

Keys piloted a Harley-Davidson built by Chet Herbert, who would later gain fame for his camshafts that he'd grind for hot rods. Herbert's bike was nicknamed *The Beast*, and it lived up to its moniker by devouring every car that it faced at the drags. Finally, Kraft figured a way to slay *The Beast*. He determined that the motorcycle was gaining the advantage at the starting line—less weight translated to quicker acceleration—so Krafty Dick (his nickname among his peers) decided to strip his Model T to its bare essentials.

Kraft showed up race day, July 30, with what amounted to a Model T frame holding a 24-stud flathead V-8 engine, a single bucket seat, a 1927 Model T cowl section to support the steering column, and a small gas tank positioned next to the driver. It was the gas tank that helped give the car its unusual name, for the can originally held insecticide for spraying bugs at his brother's 180-acre farm in the nearby hills. The car lacked a roll bar, although Krafty Dick had the foresight (with prompting by the race promoters!) to include a safety belt that he scrounged from military surplus to hold him in the seat during his sprint through the third-mile.

The car was quick, too, even though it lost its first showdown against *The Beast*. Undaunted, Kraft pressed forward with his project, and eventually eliminated *The Beast* in a subsequent early round of eliminations, allowing Kraft to stage his new concoction in the finals against a car driven by well-known driver of the time, Tom Cobb. Kraft and *The Bug* won that day, in the process becoming the first "rail job" to win Top Eliminator honors. His Top Eliminator time was 109 miles per hour, and within a few weeks it took a car capable of 120 miles per hour in the quartermile (eventually the promoters determined that to be the ideal distance) to win Top Eliminator at the Santa Ana Drags. Thanks to Kraft and his odd-looking racer, drag racing had established itself as a bona-fide form of motorsport, and today it is among the leading forms of automobile racing.

A few years ago a re-creation of *The Bug* was built for display in the Don Garlits Drag Racing Museum in Florida. It wasn't until 1998 that Kraft himself, with the help of his lifelong friend Ron Roseberry, decided to bring what parts remained from the original *Bug* for a restoration project. Six months later Kraft and Roseberry rolled this car out into the daylight.

You might say that *The Bug* survived the test of time in much the same manner as the proverbial family hammer. (It's the same hammer my grandpappy used; the handle's been changed a few times, and we had to replace the hammer head once—but it's the same hammer!) When Kraft decided to bring back *The Bug*, he journeyed to the dark corners of his storage shed, emerging with several original *Bug* items, including the tripower carburetion, early-Ford rear end, and the 1924 Model T frame. Everything else, he says, are "new parts."

Foremost, the flathead V-8 engine is not the same as the original car's. The 1950 racer had a 268-cubic-inch 24-stud motor; the current version is powered by an older 21-stud that also lacks Evans heads as on the original *Bug*. Says Kraft today, "We're looking for a good 24-stud. It's got to be 268 cubic inches, though. I had more luck with it (268-cubic-incher) than a 296." As it happened, the lightweight car—Kraft figures that *The Bug* weighed about 1,200 pounds during its glory days—couldn't, and didn't, need a whole bunch of horsepower to launch it quickly down the quartermile. In fact, the 296 easily spun the rear tires at the start, which prompted Kraft to slip in the 268. Already the young racers were gaining an education about traction at the strip.

The Bug helped Krafty Dick learn about another means of boosting horsepower, too. About that time he started fiddling with nitro fuel blends, relying on an outfit called Francisco Laboratories, located in nearby Los Angeles, to mix the magic potions that would put additional fire in the holes for all of Kraft's racers at the dry lakes and drags during those pioneering years.

Beyond the success that Kraft enjoyed with this modified Model T, *The Bug*'s true place in hot rod history remains in how it helped transcend the spirit of hot rodding to the burgeoning sport of drag racing. Kraft, who at 77 years old remains as spry now as when he raced *The Bug*, points out that his car really was nothing more than a junk-yard dog: "Everything on *The Bug* was a castoff. My brother used to have dump trucks, and he'd give me the throw away parts," he recently explained.

Yet those "throw away parts" embodied the true spirit of hot rodding, for as Kraft proved in 1950, a person with a little bit of ingenuity and a lot of resolve could show up on race day and be a winner. In Kraft's words today, "I'm the only guy who had a $200 car win a bunch of trophies against guys with legitimate race cars."

There were no specially built parts for *The Bug*. Nearly 50 years later, while reflecting on *The Bug*'s importance to hot rod history, Dick Kraft said, "I'm the only guy who had a $200 car win a bunch of trophies against guys with legitimate race cars."

A picture is worth a thousand words. In its 25th anniversary issue, *Hot Rod* magazine celebrated *The Bug*, concluding with these words: "And while today such a conglomeration of crud would be considered a bad trip, its no-nonsense design philosophy continues to be the basic foundation on which all rails are built."

Dick Williams'
1927 Ford Roadster

123 Miles Per Hour at the Dry Lakes and One Nine-Foot Trophy at Oakland

Original builder: Dick Williams
Originally built: 1952
Current owner: Blackie Gejeian

At just a glance Dick Williams' 1927 Ford roadster appears to be like most other typical early 1950s-style hot rods. Its topless Model T body sits low behind the chromed and polished flathead V-8 engine, and

Today Dick Williams' 1927 Ford roadster sparkles as brightly as when it won the nine-foot tall America's Most Beautiful Roadster trophy in 1955. Even though the car has been painted candy apple red (it originally was light blue), the car maintains its original chrome plating.

Williams showed what could be done when you based your hot rod on a custom-made frame. The hand-made frame allowed him to channel the body for an ultra-low stance, a look that became especially popular during the early- and mid-1950s.

The formed belly pan and rear rolled pan amplify the roadster's stance. The chrome wheels and wide-oval tires are not original; the 1953 AMBR winner had Ford solid wheels with narrower bias-ply tires.

of course, there's no evidence of cookie-cutter billet aluminum pieces to detract from the car's individuality.

But look closer and you'll see that the Model T hot rod's construction boasts all the right stuff to have earned it the most prestigious title presented to a hot rod in 1953 or today; this was the fourth car to win the America's Most Beautiful Roadster (AMBR) award, given to the top car at the annual Grand National Roadster Show. (For 48 years the show was held in Oakland, California, but as of 1998 the oldest car show in America has called San Francisco home.)

Williams was a young hot rodder living in Berkeley, California, when he built this roadster. His primary goal was to own a street roadster that he could readily convert to a drag or dry lakes racer. As he told the Oakland show promoters in 1953, "(The) car is primarily for street use; but (it) can be made ready for lakes, drags, or road racing in less than one hour."

The key to this four-in-one design can be found in the frame, which Williams constructed from scratch using 2 1/2-inch-diameter chrome moly tubing. By building his own frame, Williams was able to engineer practically all of the chassis dimensions necessary for a compact and efficient design. Every inch of the car is dedicated to a specific purpose. For instance, the brake and fuel lines, and the electrical wiring were routed through the left frame rail, and a pair of 9 1/2-gallon fuel tanks that Williams crafted from turnplate steel (it was the same type of material used to construct NASCAR stock car gas tanks in the early 1950s) were nestled in front of the Halibrand quick-change rear end. To give the car its low stance, Williams formed two rear "kick-ups" in the frame rails, reinforcing them with a pair of steel-plate gussets, then mounted a Model A buggy spring to clear the late-model Halibrand third-member housing.

The relocation of the gas supply from behind the firewall to the rear trunk area necessitated special fuel pumps to transport the gasoline

The interior has been reupholstered since the car won the Oakland show in 1953. Even so, the roadster's cab maintains the simplistic charm that highlighted hot rodding during the 1950s.

to the engine. Williams solved that problem by installing a pair of Carter internal pusher-type electric fuel pumps. These pumps were common among big trucks, and the pumps' moving parts are actually lubricated by the gasoline fluid that passes through them. The fuel lines lead to a quartet of Stromberg carbs atop an Edelbrock manifold that feed a 286-cubic-inch Mercury flathead engine. The Merc V-8 was ported, polished, relieved, and balanced, and spins a Howard M-16 camshaft within its "chromed crankcase," as the motor was described in the 1953 Oakland Roadster Show program.

In fact, buckets of chromium were doused over many of the car's parts. According to a feature story in the June 1953 issue of *Hop Up* magazine: "*Everything* on the front end is chromed. Spring leaves, U bolts, shackles, spring perches, axle, spindles, tie rod, wishbones, shocks, headlight brackets, drag link, pitman arm, rear support bracket for the split wishbones and all nuts and bolts," wrote *Hop Up*'s editors.

Moreover, it was quality chrome-plating. According to Blackie Gejeian, who currently owns the Williams T, this is the original

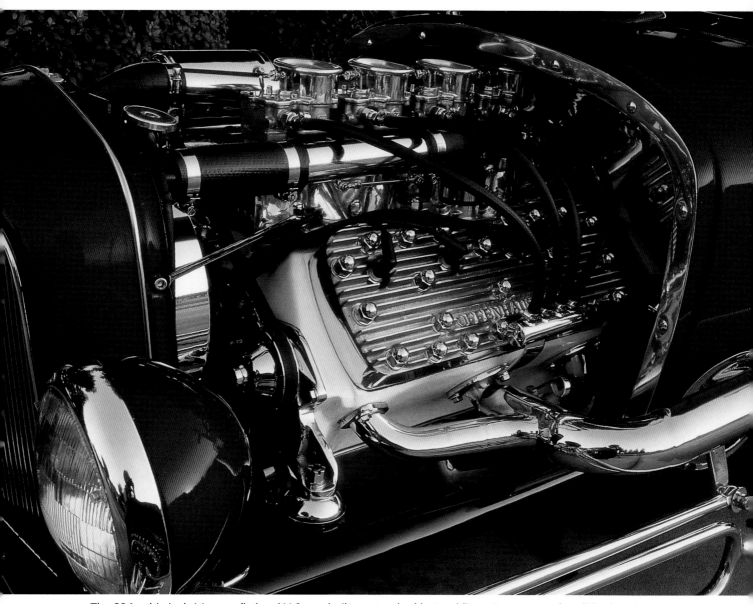

The 286-cubic-inch Mercury flathead V-8 was built on standard hot rod lines: It was ported, polished, and relieved, and a Howard M-16 camshaft spun within the smoothed-and-painted engine block. The firewall was recessed to clear the multicarb induction system.

chrome-plating that helped the car win the nine-foot-tall AMBR trophy in 1953. Blackie is quick to tell about the time that he showed the car at the 1997 Pebble Beach Concours d'Elegance. One of the show judges was about to knock off points because he thought that the old roadster had new chrome-plating when Blackie interceded. Relates Blackie: "I said, 'Excuse me, sir, but what's that on the sheet about all the new chrome?' I set him straight, and he apologized. Now, you don't find chrome-plating like that anymore."

Sadly, the paint and upholstery aren't original. In 1953 the car was light blue, and the interior was stitched in what Williams described as black leatherette. But 20 years later the car was given a new coat of paint when Dick Faulk put the car on display at the Oakland Roadster Show to commemorate its 20th anniversary as an AMBR winner; it was painted its most recent color—candy apple red—shortly before Blackie acquired the car in the early 1980s. The new upholstery is dark brown Naugahyde, although it maintains pretty much the same rolled-and-pleated pattern today as it did in 1953.

The dashboard houses fewer gauges now than in 1953, and gone is the three-spoke "roadster" steering wheel, replaced by a dished steering wheel. The tires, too, reflect a different mood than when Williams owned the car. Even so, the current rubber—Goodyear wide ovals—is about 25 years old, and could be found on the roadster's reversed chromed rims when Faulk put the car on exhibit in 1973.

It's interesting to note, too, that the Kinmont brakes have remained on the car through the years. Kinmonts were early-style disc brakes, and were considered high-tech components back in 1953 when most hot rodders relied on traditional drum/hub brake assemblies for their cars. In step with this high-performance theme, Williams fashioned the wheels using 1950 Mercury wheel centers with reversed rims from a 1950 Ford (front) and Lincoln (rear). The Lincoln rims were slightly wider, allowing him to mount big-print 8.20x15 tires on the rear. The front tires were 5.50x15.

The workmanship put into this roadster was truly superb, leading *Hop Up* to conclude its article in 1953 with the following: "The painstaking work and the attention to small details put in by Dick in the two and one-half years it took him to finish the roadster are evident from the photos accompanying the article. The thought and effort that went into this car has paid off though because the [Oakland] show judges picked Dick's roadster as being the best all around street roadster in the show."

According to the car's current owner, Blackie Gejeian, the chrome-plating is all original. Blackie tells the story about when he entered the roadster at the Pebble Beach Concours in 1997, where he had to explain to the judges that the chrome was, indeed, untouched since 1953.

Chuck Krikorian's
1929 Ford Roadster

Little Has Changed on *Emperor* Since It Was Named America's Most Beautiful Roadster in 1960

Original builder:
Chuck Krikorian/George Barris
Originally built: 1959
Current owner: Blackie Gejeian

In the book *Oakland Roadster Show: 50 Years of Hot Rods and Customs* that I coauthored with Andy Southard Jr., I wrote in chapter 2 that the *Ala Kart* was the car

What started out as a car for the local drag strip ended up as the America's Most Beautiful Roadster winner in 1960. Chuck Krikorian did much of the assembly himself, although George Barris is credited with the eye-catching body work that makes this car truly unique among hot rods of all time.

that forever changed the tone of the Grand National Roadster Show. Boasting an all-chrome undercarriage and extensive custom body work by the master himself, George Barris, *Ala Kart* raised the level of competition for the coveted nine-foot America's Most Beautiful Roadster (AMBR) trophy to new heights. In the process, the famed Model A roadster pickup became the first two-time AMBR winner, with back-to-back titles in 1958 and 1959.

By 1960, a roadster equally as beautiful and well built won the Oakland trophy. This AMBR winner was *Emperor*, another Model A roadster that also hailed from the city of Fresno, California. In fact, *Emperor* shared more than just common ground with *Ala Kart*: the two cars featured coach work by the Barris shop, and the owners of both cars were related by marriage. *Ala Kart's* Richard Peters happened to be the brother-in-law of *Emperor's* Chuck Krikorian.

Interestingly, when Krikorian set out to build his roadster into a hot rod, visions of drag racing raced through his head. He was only a high school sophomore, too young to drive, yet even then he fancied himself as a drag racer. His 1929 Ford roadster would help him achieve that goal—once he got it stripped and modified for the strip, that is. Those plans went afoul when his brother-in-law Richard Peters, and another local hot rodder of Armenian descent, one Blackie Gejeian, showed up at young Chuck's doorstep with some advice about what to do with the Model A sitting in the driveway. "They (Peters and Blackie) told me that I should build a show car out of it instead, that drag racing would cost me too much money," recalls Chuck.

The young Krikorian took the bait, hook, line and sinker, and the rest, as they say, is history. In this case, hot rod history, leading to one of the most highly awarded show cars ever built. But

Its stance nice and low, *Emperor* looks set to rule the road. From the get-go, Chuck Krikorian decided his showboat would also be a go-boat, and so he made sure that its Cadillac engine had plenty of steam.

What do hot rodders remember most about *Emperor*? Its front end! The curvaceous shell and grille insert were formed by Barris, while Krikorian was responsible for adding the oval tube fixtures. A similar treatment was added to the back of the car.

before waxing eloquent about *Emperor*'s timeless beauty, Chuck—along with Blackie, the car's current owner and Krikorian's close friend to this day—makes it quite clear that the 1929 Ford showboat was also a go-boat. Blackie tells of the time back in the mid-1960s when he set a class record with *Emperor* at nearby Kingdon Drag Strip:

"I made a 106-mile-per-hour pass," boasts Blackie, "a new record for street roadsters!" Then he mumbles something about blowing up *Emperor*'s elegant painted-and-chromed 1957 Cadillac engine. Continues Blackie, "Chuck wasn't there, and when he showed up later in the day I told him that I blew up the motor. I thought he was going to get mad or something, but you know what he says? He says, 'Who cares, we got a new record!'" Spoken like a true quarter-mile warrior. They rebuilt the motor, and to this day the Caddy V-8, with its six-pot induction system, sounds and revs crisply like a race motor. I know, I heard it myself.

A similar incident involving the 1929 beauty queen occurred at a hot rod event in Santa Barbara, California, a short time later. Chuck was standing nearby with a few of his buddies when he overheard a spectator comment about how the car—laden with chrome, candy Burgundy Tangerine paint and pearl-white upholstery—probably didn't run worth beans. A trailer queen in the highest standard, chimed the stranger.

Winning the AMBR in 1960 required some unique styling treatments. In the case of *Emperor*, bucket seats were installed in the interior, and everything was dressed with pearl-white Naugahyde.

As with the rest of the car, the Martinez-stitched interior remains the same today as when *Emperor* scooped the AMBR prize in 1960. The dashboard has speedboat gauges, and the steering wheel is from a Thunderbird. Note the compact shifter on the floor!

Wrong thing to say around Chuck Krikorian, who's proud of the way his hot rod and motorcycle engines perform. "So I jumped in the car," relates Chuck today. "I fired up the motor, then lit up the tires." Even as the tire smoke still wafted in the air and the point was clear, Chuck elected to keep the rubber burning through second gear, too. That's when all that Cad power stripped the LaSalle transmission's second gear clean of its teeth. As Chuck tells the story, "It just shredded them right off!"

Despite such heroics, *Emperor* remains pretty much intact today, and it's essentially the same car now as when it won The Trophy at Oakland in 1960. Part of its preservation can be attributed to Chuck wanting to secure the car for a later date when his son would be old enough to drive. In hopes that the next generation of Krikorians would want to enjoy his hot rod as much as he had, Chuck parked the car in the garage for what turned out to be 20 years. A farsighted man, Chuck kept the show-stopper wrapped under tarps, blankets, and other protective layers. "It was pretty much sealed from moisture and temperature change," adds Chuck.

When the day came to receive his driver's license, however, the younger Krikorian expressed little interest in the former beauty queen. Overnight, Dad had a white elephant, or so it appeared. Re-enter Blackie into the car's pageantry of color. A former winner of the coveted Oakland Roadster Show trophy (1955), Blackie had entertained the ambition of collecting the first 10 winners for a unique collection exclusive to those former AMBR cars. His collection would be a special tribute to what he considers the most important custom car show in the world. The collection's crown jewel would be *Emperor*. So he made an offer to Chuck, and soon the car—sporting the same paint, upholstery, and chrome as it did in 1960—was pulled out of mothballs, and rolled into Blackie's garage, where it sits today.

Even though Blackie couldn't fulfill his dream of gathering together the first 10 AMBRs, the fact that he acquired *Emperor* practically guaranteed that the car's magical beauty would be preserved forever. Even though the candy paint has lost much of its tangerine color tone, the old lacquer is rather well preserved, albeit with a few cracks here and there. To the untrained eye the interior looks show-ready, although Chuck is quick to point out that the padding has sagged beneath the pearl-white Naugahyde.

Perhaps most astonishing of all is the chrome. During the four years that it took to build the car, Chuck sent all of the parts that needed chrome-plating to nearby California Plating. The steering linkage, front and rear suspension, even the frame were dunked into the vats for what turned out to be high-quality chrome-plating.

No doubt, the car's most remarkable styling feature is its grille. The Barris shop installed the mesh insert, using material that Chuck scrounged from his garage. The completed grille work looked nice, but it lacked dimension. Chuck studied it for several days, then when he spotted the oval-shaped tubing that normally was used for airplane struts stacked in the corner, the idea hit: cut in cross sections, the tubing would complement the shape of the Barris-built grille shell. So Chuck got to work, cutting and fitting the oblong sections onto the grille. He gave the insert above the rear roll pan the same treatment, resulting in one of the most unique and recognizable custom lines ever given a hot rod.

Today *Emperor* holds court on occasion so that today's hot rod community can appreciate

Don't think that this AMBR-winning roadster is based on beauty alone. When the Caddy motor's six two-barrel carbs open their venturis, and the exhaust comes spitting out the chromed short-stack pipes, you realize right away this is a hot rod, pure and simple. Any doubts are erased, given the fact that *Emperor* held the street roadster class record at Kingdon Dragstrip during the early 1960s.

its beauty and the workmanship that went into building it 40 years ago. Blackie put *Emperor* on exhibit during the Oakland Museum's tribute to hot rodding in 1995, and the former AMBR winner was one of the centerpieces at the 50th anniversary of the Grand National Roadster Show in 1999.

Ala Kart may have been the hot rod that changed the America's Most Beautiful Roadster award forever, but *Emperor* certainly has its place in history, too. For no other former AMBR winner from that era is in such original condition as this car. The emperor is alive! Long live *Emperor*!

Chapter 9

Mike Haas'
1923 Ford

A History as Colorful as Its Paint Job

Original builder: Mike Haas
Originally built: 1972
Current owner: Blackie Gejeian

*B*y the end of the 1960s, hot rodders and custom car painters had developed many interesting and evocative techniques and colors unique to the automotive world. Hot rod historians attribute many of those advances to painters such as Larry Watson, Joe Bailon, Ed Roth, George Barris, Kenneth Howard (better known as Von Dutch), and other

Mod Rod's psychedelic flip-flop paint job is apparent in this photo. Although the graphics are similar on both sides, Haas painted the right side with earth tones, and used blues for the left side. The rear section is from a 1969 Corvette, grafted onto a 1923 Model T body.

individuals who weren't afraid to express themselves in search of new trends and styles.

Watson, in particular, pioneered many interesting painting techniques, and was considered the master of his time. By 1959 he became known for his glow paints, colors that offered surreal brilliance and unusual depth. Next, Watson ventured into the world of metalflakes, and later he created what he termed "flip-flop" paints, a blend that had ground-up abalone shells mixed with toners to give the paint an eerie color shift. By 1960 Watson introduced a design pattern that he called "seaweed flaming," where the licks of flame were stringier and longer than the popular flames that had been developed by Von Dutch. Other Watson-inspired painting tricks included spider-webbing and candied rainbows, among others. These and other unusual trends and applications had a profound effect on painters who were to follow, among them a young man named Art Himsl, who operated out of a small shop in Concord, California. By the end of the decade Himsl would turn out some very imaginative paint jobs

Mike Haas originally built *Mod Rod* in 1972. At that time the car was called *Odyssey*. Haas, who was a business partner with Art Himsl, built the car as an exercise in styling, applying many of the building techniques that Himsl was noted for.

Mod Rod typified the hot rod styling influences of the 1970s, including wild abstract paint jobs on T-buckets that had spindly chassis components. Excessive chrome-plating was the order of the day, too.

of his own, distinguished by their bright colors, wild graphics, and three-dimensional ribbons.

It's safe to say that Himsl officially arrived on the scene in 1969 when he and his brother Mickey built *Alien*, that year's America's Most Beautiful Roadster (AMBR) winner at the Oakland Roadster Show. Himsl's progressive painting methods not only helped sway the judges, it also grabbed the attention of Andy Brizio, another hot rod builder from the Bay Area. Brizio had been dabbling with Oakland Roadster Show entries for several years, but with little success. For 1970's show he decided to go for the gold, commissioning Himsl to give his T roadster a radical paint job. The investment paid off, and Brizio's *Instant T* drove away with the nine-foot trophy that year. Himsl had, without doubt, etched his name among the all-time great painters.

Following the success of back-to-back AMBR paint jobs, many doors opened up for Himsl. One of his business ventures in the early 1970s led to a limited partnership with another entrepreneurial hot rodder of that era, Mike Haas. Like Brizio, Haas had been especially taken by Himsl's painting skills, and he wanted to try his hand at creating some colorful muralesque paint jobs too. His odyssey into the world of wild paint jobs started with a 1923 Model T-bucket that would eventually become known as *Mod Rod*.

Haas built the car early in 1972, basing it on a 1923 Model T body. From there he went wild with the project, grafting on the rear deck of a 1969 Corvette, and fashioning the Model T's nose for a more streamlined effect.

Tony Cotta built the chassis, and Haas selected an early Chevrolet small-block engine with a Powerglide transmission for the powertrain. As with most T-buckets built back then, the steering box was from a 1956 Ford pickup truck, and the small-diameter deep-dish steering wheel was made by Covico. Ken Foster at A Action Interiors laid down the brown and orange Naugahyde upholstery, and when it came time for the paint, Himsl lent his expertise when necessary, although Haas did most of the work himself.

The *Mod Rod* graphics were applied with hologram tape. This was similar to the *Odyssey* graphics that originally were on the side of the car. Most of the paint is original, applied in 1972 by Mike Haas under Art Himsl's supervision.

According to a feature story about the car that appeared in the book *Famous Customs & Show Cars*, Haas spent about two weeks painting the modified T-bucket. When the project was finished, he dubbed the car *Odyssey*, using hologram adhesive tape to inscribe the name on both sides. *Odyssey* debuted that summer, and was voted People's Choice at the Visalia Roadster Roundup, held in September.

Later that year the car made its national debut in the December 1972 issue of *Street Rod* magazine. It was that month's cover car, and received a full color spread inside. The following year the car appeared on the cover of *Hot Rod Show World*, but by then the radical T-bucket became known as *Mod Rod*, and was one of the main draws on the East Coast show car circuit.

All show and no go? Hardly, as Charlie Gejeian takes *Mod Rod* out for a shakedown run before the 50th annual Grand National Roadster Show. The rules for the show state that the car must enter and leave under its own power.

The nosepiece was a concoction of a stock Model T grille shell and some 1970s wild-think. The mural reflects the influence that Art Himsl had on Mike Haas's first real attempt at a wild paint job.

In fact, according to current owner Blackie Gejeian, throughout the remainder of the 1970s, *Mod Rod* logged more miles on the show circuit than any other car.

When it was retired from the show fleet, *Mod Rod* was auctioned off, and eventually fell into the hands of Jack Quayle, who purchased it as a sales gimmick for his auto repair business in Northern California. Quayle's original idea was to park *Mod Rod* in front of his shop to attract business, but he found himself driving the car more than parking it as a promotional oddity.

"I drove it all over Northern California," Quayle said recently. "That car's a train. It never overheated; it was a great-running little car."

Quayle could have added that it was slippery-quick, too. One time at Tom Pufer's annual hot rod gig at Fremont Raceway, *Mod Rod* posted a top speed through the quartermile at 103 miles per hour, with an elapsed time in the high 13s!

Quayle tells another interesting story, this one about when driving to Andy's (Brizio) Picnic. *Mod Rod* got a flat tire, and Quayle had to leave the car on the side of the road with a friend while he took the deflated tire to a nearby gas station for repair. When he returned, there were four busloads of Japanese tourists parked nearby. "Every one of those tourists had a camera," recalled Quayle, "and they were all taking turns taking pictures of each other with the car. It was funniest thing I ever saw."

As for the name change from *Odyssey* to *Mod Rod*, neither Quayle nor Blackie can account for how or why it came about. Quayle said that the car was billed in *Hot Rod Show World* as the car driven by Michael Cole's character in the television series *Mod Squad*. "People came up to me and said that they remember seeing the car on the TV show," said Quayle, "but I've seen lots of them (episodes), and I never saw it."

Regardless of its Hollywood past, *Mod Rod*'s major contribution to rodding today can be found in its paint job. The wild outer space scene that wraps around the whimsically styled body is a rolling example of an era when expressive young men bombarded the hot rod world with psychedelic paint jobs highlighted by wild graphics and creative murals. There may never be a time like it again in hot rodding, but thanks to survivors such as *Mod Rod*, those days will burn forever in the hearts of hot rodders everywhere.

Chapter 10

Dan Woods'
1917 Ford Roadster

IRS without the Audit

Original builder: Dan Woods
Originally built: "About 1972"
Current owner: Blackie Gejeian

*H*ot rods have always been considered an American phenomena, incorporating an ensemble of made-in-the-U.S.A. parts with domestic cars. Indeed, the phrase "foreign car parts" is practically foreign to the American rodder's jingoistic pride. Yet back in the late 1960s and 1970s some American hot rod builders found favor with a certain British-made

Dan Woods built this T-bucket for Bill Block nearly three decades ago. T-buckets were growing in popularity about this time, and Woods was considered a master at building them. Some of his specially built cars sold for $50,000 or more. Their presence helped usher in the age of the high-dollar customer cars.

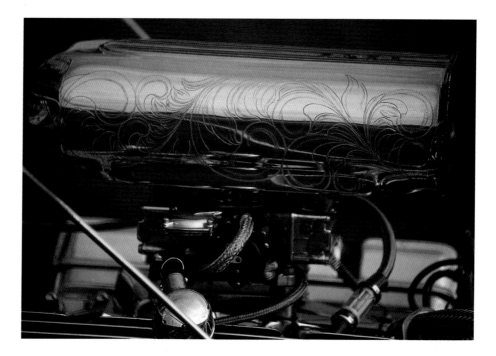

Some of the expense for customer cars was absorbed in the detail. Here the chrome-plated carburetor air scoop shows off some rather ornate engraving. Note, too, the brass plating and braided aircraft hydraulic lines.

rear end. We're talking about the Jaguar XKE's independent rear suspension (IRS), a subassembly that incorporated double coil-over shock absorbers, inboard disc brakes, short uprights, and stubby half-shafts for a compact design. Hot rodders realized the advantages the XKE IRS offered, and practically overnight its popularity grew. The Jag rear ends were plentiful and cheap, too; all it took to acquire one was a few dollars and a quick trip to the local junk yard.

Joe Cordoza is credited as being among the first to hang an XKE rear assembly onto a hot rod. Eventually, though, many of the top-name builders joined the revolution, switching to the English-made rear end. And it was during the early part of the 1970s that a certain Southern Californian named Dan Woods gained a reputation as *the* man to see if you wanted a hot rod with an XKE rear end.

Woods represented the carefree spirit that embodied the hot rod industry at the time. A charter member of the Early Times Car Club, Woods liked nothing better than to have Fun (with a capital F!) with hot rods. "We got a little crazy back in those days," Woods recalled of the early times with the Early Times. "But it was all good-natured, harmless fun, really."

No matter, because when it came to building hot rods nobody was more serious than Woods, and his shop in Paramount, California, turned out some beauties. Today he points out that it wasn't unusual for a customer to pay in the neighborhood of $50,000 for a Woods-built hot rod. In 1970s dollars, that was a lot of money, akin to posting a check today with a six-digit figure crowding the amount column.

Woods' calling card was a Ford T-bucket with a Jag rear end. Woods (who today custom builds steel homes worth millions of dollars each) says that he built this particular T-bucket using a 1917 Ford body ". . . in about 1972, it's been so long I really can't remember." The car was built for Bill Block, who happened to own Cragar Industries at the time.

Originally Woods powered the 1917 T-bucket with a small-block Ford that breathed through a quartet of Weber carburetors and rare Gurney-Weslake heads. Woods was a close acquaintance of race car builder Dan Gurney, who passed along the race-bred components for the project car. Unfortunately, the engine was discarded sometime during the next 20 years, replaced by the late-model aluminum Buick that currently sits within the frame rails.

Woods also says that the car originally had one of the first digital-gauge dashboards in a hot rod. There were no digital-gauge kits on the market in 1972, so Woods built the system from scratch. Sadly, the unique instrument cluster, which had a series of five rectangular-faced gauges, has since been removed from the car, replaced by a set of analog dials in the burl wood dashboard. Another avant-garde design feature was the car's firewall, milled from aluminum stock. A decade later the entire hot rod industry would discover billet aluminum as a source to fill the need for any and all aftermarket requirements, but in 1972 billet aluminum for custom treatments was a novelty.

Sometime during the past couple of decades the Naugahyde upholstery was stitched over with fabric, and a few other odds and ends have been replaced, too. But for the most part the car is the same as when Woods completed it twentysomething years ago for Bill Block. That includes the beautiful aluminum gas tank that master-metalsmith Steve Davis formed when the car was first built, the 1909 Ford brass headlights, and the tube front axle with lay-down coil-over shocks.

Eventually the T-bucket found its way into the hands of Mike Kazian, who babied it with

Woods commissioned ace metal fabricator Steve Davis to construct the aluminum gas tank. It was work such as this that brought hot rodding to a new plateau in which every little detail of the car had to be spot-on perfect.

The interior has been changed extensively from the original. Gone are the five rectangular digital gauges, replaced with more conventional round analog dials, and the upholstery has been restitched with fabric and vinyl. Note the air horns, which were popular on T-buckets during their heyday.

The trademark of a Dan Woods-built T-bucket became the Jaguar XKE rear end. Of this particular assembly, Woods said, "I massaged the third member a whole bunch. It's all smoothed and chromed. It was 20 years ahead of its time."

TLC until he passed away in 1994. Kazian's close friend Blackie Gejeian acquired the car from the estate, and immediately had a plan: "I knew that the 50th anniversary of Oakland (Grand National Roadster Show) was coming up, and I wanted to enter this car. It's a Dan Woods-built car, and it's one of the finest with a Jag rear end," explained Blackie, who also promotes his own car shows in Fresno, California.

To ready the car for the gala 50th annual Grand National Roadster Show in 1999, Blackie disassembled the T-bucket's Jag rear, and "chrome- and brass-plated everything that needed it." Mission accomplished, he readied the car for the show, and decided that the entry would be under his late friend's name. "I entered the car in honor of my good friend, Mike Kazian," said Blackie. He also paid homage to the car's original builder: "I wanted Woods to be the builder on record. He hasn't been given the credit he deserves in hot rodding. Woods did beautiful work."

Today when the name Dan Woods comes up now and again in conversation among hot rodders, the first thing they make reference to are the chromed-and-polished Jag XKE independent rear suspension assemblies that identified the Southern California builder. Fittingly, when I contacted Woods about this T-bucket, the first thing he recalled was the old Ford's IRS. Said Woods, "I massaged the third member a whole bunch. It's all smoothed and chromed. It was 20 years ahead of its time."

Perhaps the Jag IRS on this car was ahead of its time, but beyond the car's many unique features, it was—plain and simple—built right and tight. In Woods' words, "That's the best T-bucket I ever built." Quite an endorsement from a man who says that he built about 160 hot rods during the 18 years his shop was open for business.

The front view shows off the intricacy and symmetry of the car, right down to the opposed lay-down coil-over shocks. The brass and gold plating remains in exceptionally good condition.

Chapter 11

Lil' John Buttera's
1926 Ford Sedan

A Tall T That Filled a Tall Order

Original Builder: John Buttera
Originally built: 1974
Current Owner: Mike and Jo Sweeney

Few people can, or will, argue that the 1970s was a decade like no other when it came to tastes in style and fashions. Unisex clothing originated during the 1970s when women in the professional workforce began to include pants suits as part

Lil' John Buttera mastered his building techniques constructing funny cars and top fuel dragsters. He built this 1926 Model T sedan as part of a series that Terry Cook wrote for *Hot Rod* magazine, entitled the "Great California Street Rod Civil War." Buttera became a celebrity among the street-rod crowd overnight.

of their daily dress code, and casual attire for men focused on hip-hugger flare-bottom jeans and perm-curled hairdos. In the world of music the 1970s gave us acid rock and disco, while a knee-slapping television musical variety show known as *Hee Haw* rose to the top of the rating charts, joined by other television favorites *including The Brady Bunch, Happy Days, Welcome Back Kotter, Starsky and Hutch,* and *The Donny and Marie Show.*

Closer to home, the hot rod world didn't escape the surreal 1970s, either. Today, 1970s refugee street rods stand out like rolling paint blems; trendy color schemes from the 1970s include two-tone metallics, ribbons and scallops, and spiderweb overlays. And if the psychedelic paint job doesn't leave a clue to a particular hot rod's vintage, then perhaps its chassis works will. In vogue twentysomething years ago were chromed wire-spoke wheels that were

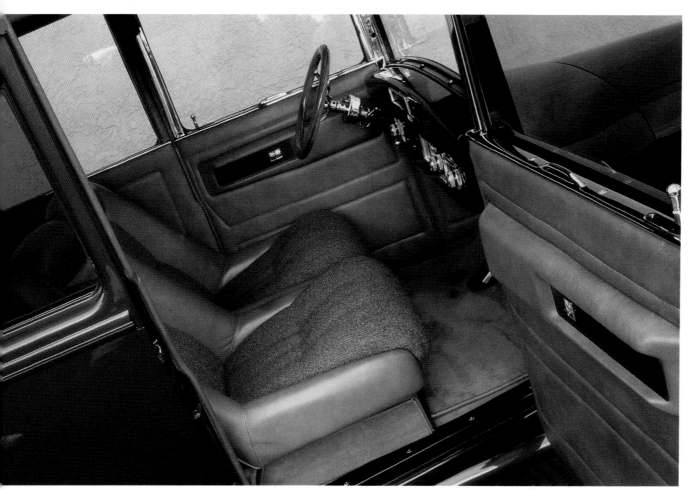

Through selective use of leather, tweed fabric, polished aluminum, and chromed parts, Buttera showed the hot rod world how you can make a silk purse from a sow's ear. Open the door and its plush styling beckons you to jump in and go for a drive!

wrapped with raised white-lettered wide-oval tires, and stock-height I-beam and tube front axles were matched with protruding wide rear axles. Under the hood you might even spot a four-cylinder Ford Pinto engine, or a V-6 of questionable origin. Many enthusiasts today feel that hot rodding experienced its dark ages during the 1970s.

It took a car the likes of Lil' John Buttera's 1926 Ford tall T sedan to help shake hot rodding out of its 1970s slumber, for this hot rod not only boasted discriminating taste in style, it also broke new ground in how a hot rod should be built. Beyond its appealing looks, the Buttera T was founded on superb workmanship by its builder, leading Gray Baskerville from *Rod & Custom* magazine to proclaim in a drive-test article that he wrote about the sedan 21 years after it was built: "... now I know what it's like to go for a spin in what just may be the finest street rod ever built."

What may be the "finest street rod ever built" didn't happen overnight, though. Buttera, who cut his teeth in the trade building funny cars and top fuel dragsters, admitted that he had been planning the T sedan in his head ever since he was a 14-year-old high school kid hooked on hot rods. As a grown man, he once said: "I've wanted one (Ford hot rod sedan) since I saw Art Chrisman's chopped '28 sedan."

The dream advanced to reality in 1974 when *Hot Rod* magazine summoned Buttera to partake in what *HRM*'s editors proclaimed "The Great California Street Rod Civil War." The rod buildup story pitted Lil' John from Southern California against NorCal-based Andy Brizio (who built a fiberglass-bodied 1932 Ford roadster as his contribution to the friendly rivalry story). Buttera spent about six months laboring in his shop to complete the Model T, and for the most part, this car was crafted one piece at a time, using parts that were handmade by Buttera or other hot rod specialists. Furthermore, the gennie steel body was fitted and welded together so that it would be a solid unit, then it was permanently attached to a custom-built frame.

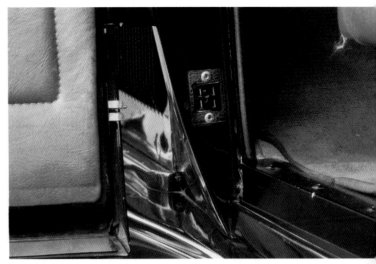

Hot rodders have always been creative with their building techniques, and Buttera was no exception. With power-everything in the interior, he had to find a way to route the wiring in a presentable fashion. His solution for the power to the doors was to install pronged connectors in the door jambs.

About the frame: Buttera wanted the sedan to offer its occupants a smooth, firm ride, so he fabricated a two-tier frame similar to what he was familiar with from his drag racing experiences. The upper rails were made of rectangular tubing, and the lowers were formed using 1 1/2-inch-diameter round-steel tubing. He then hung independent suspension at all four corners, whittling the spindle uprights out of billet aluminum, and forming the A-arms from 4130 chrome-moly tubing. The Jaguar third-member is supported by a pair of highly modified Richards uprights, and Lil' John fabricated the control arms, radius rods, and hub carriers in his typical from-scratch manner. Koni shocks cushion the ride at all four corners.

Some other big-name hot rod builders played key roles in the car's genesis, too. Metal wizard Steve Davis formed the two 12-gallon aluminum saddle tanks that fit snugly under the Model T's side aprons (Davis helped fabricate many of the other fixtures as well), and Art Chrisman rebuilt the 289-cubic-inch Ford small-block motor. And when the car was ready for upholstery, master stitchman Tony Nancy was called on to dress the Volvo bucket seats with fabric matching the body's metallic brown paint (recently repainted by Fat Jack Robinson).

Before the paint was applied, though, Buttera had a few more tricks up his welder's sleeve that would ensure there would be no annoying squeaks and rattles emanating from the car while driving on the road. To that end he gave the car what best could be described as a unibody design. First, he welded all the body seams, then he attached a single-piece roof insert made of sheet aluminum. Next he formed a one-piece floor out of 6061-T6 aluminum, attaching it, as he did the roof, with rivets spaced two inches apart throughout. Finally, the interior was ready for the amenities that Lil' John felt were necessary to make this truly a roadworthy daily driver. Upon completion, he drove the car to the NSRA Nationals in Minneapolis, Minnesota, and later his wife, Joanne, used the tall T as her commuter special until 1982 when Ray Johnson bought it. More on that later.

The creature comforts that Buttera built into the sedan's crew quarters are common fixtures among rods today, but for a mid-1970s hot rod they were truly avant-garde. His design included heater/air conditioning ducts routed directly to the dash, steering-mounted cruise

Since 1974 this T sedan has had only one powerplant, a 1967 Ford small-block engine. Art Chrisman built the original, and he was called on for the rebuild when third owner Larry Johnson decided the 289-cubic-inch motor needed to be freshened up.

Buttera adapted the Jaguar XKE rear end, fabricating many of the fixtures himself. The control arms, uprights, and routing of the exhaust tips above the half-shafts reflected many of the building techniques that Buttera had mastered while fabricating racing cars during the 1960s.

control, digital gauges, and special bulkheads to mount the bucket seats.

Second-owner Ray Johnson owned the Buttera T for about a year, then he sold it to Larry Johnson, who was responsible for the car's rehab. According to Larry Johnson, the nine-year-old car had 18,000 miles on the clock, and he wanted to freshen it up to look like new, so he tore it down to the bare frame. Also involved in the rebuild was Chrisman, who tightened up the Ford small-block engine one more time.

The car ultimately changed hands again time, and it's currently owned by Mike and Jo Sweeney, of Orange, California. The Sweeneys aren't reluctant to drive the car, either, and like Joanne Buttera, Jo isn't shy about taking the car to the local grocery store to fetch some daily staples for the Sweeney household.

Mike figures they drive their classic hot rod about 10,000 miles a year. One summer they drove it to the NSRA Nationals in Louisville, Kentucky, and it served as their ride to Mike's Midwest high school reunion in 1998. Along the way they stopped at the Brag & Drag at Bandimere Raceway near Denver, Colorado, where they copped Long Distance Award.

No doubt the Buttera T will go down in hot rod history as a true milestone car. But beyond its relevance to rodding, this tall T stands high among its peers simply because it's such a wonderful hot rod. As Baskerville wrote in his drive-test back in 1995, "You keep saying to yourself, 'I can't believe that this thing was started and finished in just six months . . . 21 years ago.'" Believe it.

The independent front suspension was engineered by Buttera, too. The purposeful design included handmade control arms and uprights, coil-over shocks, and disc brakes. Note the brackets that mount to the lower control arms for the braided brake lines.

Chapter 12

Tom McMullen's
1932 Ford Roadster

Fourth Edition for the Fourth Estate

Original builder: Lobeck's V-8 Shop
Originally built: 1997
Current owner: *Street Rodder Magazine*

Should there be credence to the old saying, "If they liked it once, they'll love it a second time," then the hot rod community must be thoroughly enraptured with Tom McMullen's 1932 Roadster, because *four* of the flamed Deuce highboys have been built since

No doubt about it, this roadster is a real head-turner. Tom McMullen built his first supercharged Deuce highboy in 1962, and the car appeared on the April 1963 cover of *Hot Rod* magazine. He built a clone in the early 1970s as a project car for his own magazine, *Street Rodder*, and since then the staff has built two more, one as a give-away car, and this one as the staff's ride.

This is pretty much the view that Tom McMullen enjoyed back in 1962, and this is what the *Street Rodder* editors look at today. Guy Shively was given the task of duplicating the flamed roadster's intricate pinstriping for Clone No. 3.

1962. The original—which is still on the road, although other than the Moon pressure tank that's still mounted between the front frame horns, you wouldn't recognize it—was featured on the April 1963 cover of *Hot Rod* magazine, and nearly 15 years later McMullen Roadster No. 2 repeated cover status, this time for *Street Rodder* magazine, a publication that McMullen himself founded in 1972.

Actually, the second car was a magazine project that Tom undertook to illustrate how he would have built his famous flamed highboy during the 1970s, when street rodding was enjoying its renaissance among enthusiasts. That car differed from the original in several distinguishable ways: it had chromed wire spoked wheels rather than unpolished American Torq Thrusts; the interior was trimmed in black vinyl instead of white and black Naugahyde; additional flames licked the roadster body's rear flanks; the front suspension was upgraded to 1970s specs; and, above all, the clone roadster lacked the original's classic trademark—the Moon pressure tank up front.

The third edition hit the road more than 10 years after No. 2, again as a project car spearheaded by McMullen. The car was built to intentionally mimic the 1963 version, and as Eric Geisert wrote in a recent *Street Rodder* article about the three clones, the third car (second clone) was "an attempt by Tom to recapture the fun of the first roadster." That highboy remained in Tom's stable until his untimely death in an airplane crash in 1995. Clone No. 2 was passed on to Tom's close friend Al King, who still owns it today.

McMullen Roadster No. 4—Clone No. 3, if you will—was built in 1997 to commemorate the magazine's 25th anniversary. As events unfolded, the clone turned out to be more than just a salute to a very famous car. The familiar 1932 Deuce highboy has become an institution at the magazine, and to this day Clone No. 3 serves full-time duty as the editors' staff car. "We take turns driving it home and back to work," explained Brian Brennan, current editor at *Street Rodder*.

From the get-go the editors decided that the fourth edition would be, to date, the most accurate reproduction of Tom's original highboy. Old photos of the original car were located, and to maintain authenticity staffers interviewed some of the people who worked on the original. For instance, Darryl "Whitey" Morgan, who stitched the first three McMullen Roadster interiors, was asked to dig deep into his memory (and photo file) to recall exactly how he did the rolls and pleats back in 1963. And, although Ed "Big Daddy" Roth wasn't called upon to pinstripe No. 4, Guy Shively was instructed to guide the horsehair with the exact same strokes on the new Wescott fiberglass body as Big Daddy applied to the all-steel original.

The actual construction of No. 4 took place at Lobeck's V-8 Shop in Cleveland, Ohio. Barry Lobeck's crew started with a pair of Just A Hobby

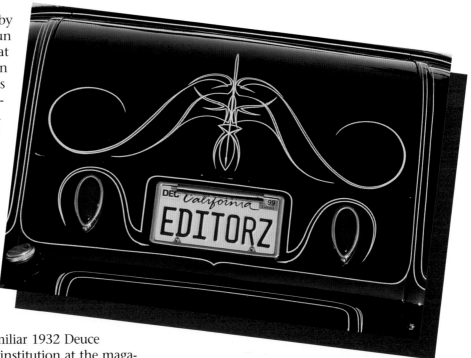

The personalized license plate says it all. This is the highboy roadster that the editorial staff at *Street Rodder* magazine share as their daily driver. Editor Brian Brennan says that the cleaning chores are left to the last person who drives it. It's a tough job, but darn it, somebody's got to do it!

frame rails, then gave the chassis only those upgrades they felt necessary for a safe ride; the makeshift improvements include a Mullins steering system, new Magnum Axle spindles, king pins, steering arm, and tie rods, things like that. In the interest of safety and authenticity, Lobeck's strengthened the 1940 Ford drilled split wishbones by fitting each lightening hole with a cross tube that helps maintain torsional rigidity for the traditional-style links.

Visually, though, McMullen Roadster No. 4 assumes the same character and stance as the car that Tom built in 1963. The Chevrolet 350-cubic-inch engine is fed by a twin-carb 6-71 supercharger (the original had a GMC blower, the clone gets its huff from a Weiand), a Halibrand quick-change rear end was fitted to the drivetrain as per the original, and the stock-appearing hood (actually a new Rootlieb reproduction) was dressed with the same flame job that Tom personally painted on the first car. Finally, even though the traditional-style roadster rides on new BF Goodrich radials (again, for safety reasons), the wheels are reproduction American Torq Thrusts.

After Clone No. 3 was finished in early 1997, it debuted at the NSRA Nationals East in York, Pennsylvania. The *Street Rodder* staff entered the familiar roadster at other events during the year as part of the magazine's 25th anniversary, and when its tour of duty was complete, Clone No. 3 reported back at *SRM* headquarters in Placentia, California, to serve full time as the magazine's official company car.

Today Clone No. 3 can be seen on the streets of Southern California, or as the flagship for the magazine's annual Tom's Fun Run. As you might guess, the editors are having a blast with the roadster. And while the editors all share in the fun, they also share in the car's upkeep. In editor Brennan's words, "The last person to drive the car is the last person to wash it, too!" Ah, the life of a magazine scribe. It's a dirty job, but darn it all, somebody's got to do it!

If you ask any old-time hot rodder what he remembers most about the McMullen roadster, chances are he'll say the flames and the Moon pressure tank mounted to the front crossbar. When McMullen built the first car, the tank was used expressly for racing. The car was good for 151 miles per hour at the dry lakes, 127 miles per hour/11.59 seconds in the quartermile.

Like the original, the small-block Chevy on car No. 4 is fed by a 6-71 supercharger with two four-barrel carbs. The new engine was given an alternator, plus the wiring and plumbing are cleaner than the original's. But the spirit of the car is unchanged!

Jim "Jake" Jacobs'
1928 Ford Touring

Reviving the Fun Factor

Original builder: Jim "Jake" Jacobs
Originally built: 1987
Current owner: Jim "Jake" Jacobs

What's an Archie-and-Jughead jalopy doing in a book about milestone hot rods, you ask? After all, any bonehead can see that this old 1928 Ford phaeton isn't—or ever was, for that matter—on the cutting edge of hot rod fashion. Look at the car: the touring tub's interior could just as well have been

Okay, it's time to get serious about our hot rod fun. Enter *Jakeopage*, a 1928 Ford touring tub that bubbles over with fun and adventure. Jim "Jake" Jacobs built the car in a matter of weeks back in 1987. Purpose of the mission: why, to have fun, of course!

stitched for *Sanford & Son*'s junkyard hauler, and oxidation more than anything else is responsible for preserving what few chrome parts were used in the project. We don't even have to mention the phaeton's paint, although we might as well be up front and state right now that the bright (?!) red base coat was applied with a *handbrush* by owner and builder (with a little help from his friends, too) Jim "Jake" Jacobs. The montage of old hot rod magazine pages plastered on the body panels? Well, that's more of Jake's doing.

Indeed, to know this car you must also know its owner, who has been a part of the hot rodding scene since the 1960s. To be sure, Jake's background is varied and colorful: he's a charter member of the Early Times (a hot rod club that was based on the concept of having fun first, being cool second); he's an ace mechanic and fabricator; and Jake was associate editor for *Rod & Custom* magazine for a few years until he joined forces in 1974 with another hot rod icon—Pete Chapouris—to form Pete & Jake's Hot Rod Parts. After Pete and Jake sold their burgeoning business in 1986, Jake took a brief hiatus—actually, more of a sabbatical—so that he could determine in exactly what direction he wanted to resume his journey through the world of rodding.

Finally, he realized that the path he wanted to take—that he *needed* to take—headed toward

The backside of the tub lets anybody who's approaching know that they are entering the fun zone. Notice, too, that Jake didn't knock himself out with smoothing the all-steel tub body, either. Another advantage to the rustic (rusting?) body panels is that Jake never worries about where he parks his fun car.

that far-off place known as *fun*. The journey began in his garage. One day when Jake, ever the innovator, ventured into the garage's nether regions, he realized that there was a wealth of old Ford parts from past projects lying about. The brain shifted gears, prompting Jake to another realization: "With a few more parts I could build a hot rod from all this," he told himself. The quest began, and Jake-the-hot-rodder embarked on a mission to gather *only* genuine Ford parts for his new project (although power comes from an early Chevrolet 283 V-8 motor).

He didn't make a big issue about it at the time (nor does he to this day), but Jake was a little disturbed by what was happening within the hot rod fraternity. By the middle of the 1980s—as Jake saw it, anyway—hot rodding was transforming into a game of high-dollar one-upmanship among many key players. In the process some owners, whose cars perhaps lacked some of the "perfection" that was otherwise found in many of the top-flight hot rods being built at the time, felt slighted, even embarrassed, by their rides. Consequently, some really nice—but not necessarily "cover-car quality" hot rods—were being left indoors, not to be seen or admired by the remainder of the rodding clan simply because their owners were ashamed that their rides weren't alpha-male material.

So Jake set about building a car that played on the fun factor, not the money factor. Jake explains, in simple terms, today: "I just felt like slamming something together so I could go out and have fun. It's sort of a fifties-style jalopy." Once the parts had been gathered under one roof—Jake's garage—he spent about a month piecing the puzzle together. Finally, during the summer of 1987, the car was ready for its maiden trip, and with 12 miles showing on the rusted odometer he pointed the car north from his Temple City, California, home. Next stop: Pleasanton, California, for the Goodguys West Coast Nationals.

The Pleasanton event was a memorable occasion for Jake's tub, too, for this was the day

A close-up of the grille shell accents the brushed-on paint job. Jake and his buddies painted the car at the 1987 Goodguys West Coast Nationals. Jake said nonchalantly, "While other people were waxing their cars, I was painting mine." Careful, Jake, don't drip paint on the pitted chrome!

that he would paint the car. A quick stop at the paint shop for a couple of buckets of Sherwin-Williams red, then it was onward again to the Pleasanton Fairgrounds where he and a few other rodding buddies donned their painters hats (true), and commenced to dipping their brushes. Recalls Jake of that milestone day: "While other people were waxing their cars, I was painting mine." A few weeks later at the annual California Roadster Show, the car scooped the annual Stroker McGurk Award, presented to the "Best Highboy" that exemplifies the spirit of Tom Medley's lovable fictitious cartoon character of the 1950s.

A few years after that Jake got another inspiration, this time while pounding the pavement of Pasadena, California's, unique downtown district, which also happens to be home for sidewalk musicians who play for handouts and fun (there's that word again). Jake noticed that many of the players had

Extra, extra, read all about it! And you can curbside with *Jakeopage*. The old hot rod magazine clippings were donated by fun-time hot rodder Tony Thacker, who had sacrificed some old magazines in the name of research for a book that he penned shortly before Jake decided to decoupage his touring tub.

pasted their musical instrument cases with travel stickers of places they'd been. Jake mumbled something to the effect of wanting to give a similar treatment to his hot rod, and his comments were overheard by close friend and all-around fun guy Tony Thacker. Tony, an expatriated Englishman who prefers highboys and roadsters any day over crumpets and tea, told Jake that he had a bunch of old hot rod magazines that he had cut up while doing research for a recent book project. The mags were going to get tossed in the trash anyway,

so what the heck, he told Jake, "Let's put them on the tub!"

So, with the help of Tony's girlfriend, Kathy Berghoff, the three set out to decoupage the Model A. Thus was born the *Jakeopage*, a car that's truly a piece of rolling hot rod history. Adds Jake, "At events people would walk up to the car and start reading the captions. You'd be surprised at how many of today's hot rodders are unaware of how it (hot rodding) all began."

Furthermore, many of those same hot rodders might not fully realize how much fun hot rodding served up to its participants during the early years. Worse yet, some of the current crop of rodders don't fully appreciate how much fun hot rodding can be today. That's where cars such as the *Jakeopage* play important roles. As Jake put it, "We've got to put a little fun and craziness back into this (hot rodding)." And since 1987 the *Jakeopage* has done just that, for its owner and the thousands of people who have been fortunate enough to experience it at rod runs that Jake has frequented with his decorated tub.

So, as we close the final chapter for *Hot Rod Milestones*, we can turn to the *Jakeopage* and say, "Let the fun begin!"

Jake gives fair warning about the speed potential of this beauty. So far, nobody has violated the dictum because, a) there are no windows in this car, and b) it has yet to top the magic one-two-oh mark.

The cab won't exactly win Best Interior at a car show, but who cares? When Jake's on the road, he simply tosses empty bags of chips and spent soda cans in the back seat or on the floor. Amenities include a heater (beneath the dash), full instrumentation, a warning sign, and power steering (old-time wheel knob).

Index

A Woman of Distinction, 29
A-V8 roadster, 25, 28
Ala Kart, 61, 62, 65
Alexander, Steve, 32
Alien, 69
America's Most Beautiful Roadster (AMBR), 13, 37, 43, 44, 57, 61, 62, 69
Arias, Nick, 26
Athan, John, 24, 26
Bailon, Joe, 67
Bakersfield Raceway, 34
Bandimere Raceway, 83
Barris, George, 61, 62, 67
Baskerville, Gray, 81
Bastian, Art, 15
Becker, Hank, 14, 20, 22
Berghoff, Kathy, 94
Best Engine award, 14
Best Highboy, 93
Bill NeiKamp roadster, 43
Block, Bill, 73–75
Brink, Dylmer, 46
Brizio, Andy, 69, 81
Brock, Ray, 39
Buttera 1926 Ford Model T sedan, 16, 81, 83
Buttera, Joanne, 16, 83
Buttera, John, 16, 79
California Roadster Show, 93
Chapouris, Pete, 41, 92
Chrisman, Art, 82
Chrysler, 28
Classic Hot Rod, 11
Clone No. 3, 88
Cobb, Tom, 52
Coddington, Boyd, 16
Cole, Michael, 71
Conforth, Mark, 15
Cook, Ray, 17
Cook, Terry, 79
Cordoza, Joe, 74
Cotta, Tony, 69
Cucaracha, 31
Davis, Steve, 75, 81
Deuce highboy, 37, 85
Deuce roadster, 39
Doane Spencer roadster, 38, 41
Don Garlits Drag Racing Museum, 52
Downey Brothers' 1926 Ford Model T, 13, 14, 19

Dry lakes speed trials, 39
East, Neal, 41
El Mirage Dry Lake speed trials, 26, 45
Emperor, 9, 14, 16, 61–65
Famous Customs & Show Cars, 70
Faulk, Dick, 59
Foster, Ken, 69
Fox, Dick, 17
Fremont Raceway, 71
Ganahl, Pat, 34
Gejeian, Blackie, 9, 11, 55, 57, 59, 61, 62, 67, 71, 73
Gemsa, Joe, 22
Gilmore, 22
Goodguys West Coast Nationals, 93
Grand National Roadster Show, 13, 16, 39, 44, 47, 57, 59, 62, 65, 69, 76
Gurney, Dan, 75
Haas, Mike, 67–71
Harley-Davidson, 52
Himsl, Art, 68, 70, 71
Honda, Bill, 22
Hood, George, 22
Hot Rod of the Month, 32
Hot Rod Show World, 70, 71
Howard, Kenneth, 67, 68
I Want You, 29
Instant T, 69
ISCA Grand National Championship, 14
Iskenderian, Ed, 26, 28, 29, 31
Jackson, Kong, 26
Jacobs, Jim "Jake," 16, 43, 91, 92
Jakeopage, 17, 94, 95
Johansen, Harold, 26
Johnson, Larry, 83
Johnson, Ray, 82, 83
Kazian, Mike, 75, 76
King, Al, 87
King, the, 25–29
Kingdon Drag Strip, 63
Kraft, Dick, 15, 49–52
Krikorian, Chuck, 9, 14, 61, 64
L.A. Roadster Show, 34
LaCarrera-Panamericana Road Race, 39
Leonardo, Tom, Jr., 26
Leonardo, Tom, Sr., 28
Little Candy Pearl, 14
Lobeck, Barry, 16

Loving You, 25, 29
Martin, Ed, 14
Martin, Karen, 14
McCoy, Bob, 17
McMullen Roadster No. 2, 86
McMullen Roadster No. 4, 87
McMullen roadster, 87, 88
McMullen, Tom, 13, 85, 86
Medley, Tom, 93
Meyer, Bruce, 37, 41
Miller, Ak, 39
Mod Rod, 67–71
Models
Deuce highboy, 87
Ford Roadster, 13, 37, 55, 81
Ford touring tub, 17
Ford, 15, 17, 31, 67
Model A roadster, 14, 25, 62
Model A, 25, 45, 57, 62, 94
Model T, 19, 22, 23, 32, 49, 50, 52, 53, 67, 81
Model T roadster, 31, 44, 61, 62
Model T sedan, 16, 19, 49, 79, 81, 83
Model T-bucket, 51, 69, 69, 70, 73–74, 76–77
Roadster, 85
Modern hot rods, 16
Morgan, Darryl "Whitey," 87
Muroc Dry Lake, 28
Nancy, Tony, 82
National Hot Rod Association (NHRA), 29, 50
National Roadster Show, 45, 47
National Street Rod Association (NSRA) Street Rod Nationals, 41, 82-83
NeiKamp Roadster, 44, 45, 47
NeiKamp, Bill, 43
NHRA California Hot Rod Reunion, 34
NHRA Motorsports Museum, 11, 29, 32
Normile, Mike, 39
NSRA Nationals East, 88
Oddo, Frank, 28
Odyssey, 68, 70
Paramount Studios, 29
Pasedena Reliability Run, 45
Pebble Beach Concours d'Elegance, 37, 41, 46, 47, 59
Peters, Richard, 62

Petersen Automotive Museum, 41
Po, Bob, 29
Presley, Elvis, 25, 27, 29
Quayle, Jack, 71
Rail dragsters, 49
Road Rebels, 28
Robinson, Fat Jack, 82
Root, Mike, 17
Roseberry, Ron, 52
Roth, Ed "Big Daddy," 67, 87
Russell, Art, 14
Russell, Dick, 46
Santa Ana Drags, 46, 50–52
SCTA (Southern California Timing Association), 13
Seiden, Gary, 15
Seiden, Richard, 15
Senter, Louie, 26
Shively, Guy, 86, 87
SOHC Model T engine, 23
Soup job, 19, 20, 51
Southern California dry lake bed, 31
Spencer, Doane, 37
Stroker McGurk Award, 93
Suicide front end, 21
Super job, 19
Sweeney, Jo, 79, 83
Sweeney, Mike, 79, 83
T roadster, 35, 69
Thacker, Tony, 94
The Beast, 50, 52
The Bug, 49, 50, 53
The Elvis Car, 25, 27–29
The Great California Street Rod Civil War, 81
Thelan, Don, 16
Thomas, Larve, 22
Timing tags, 45
Tito, 29
Tom McMullen roadster, 15, 16
Top Eliminator, 51
Tudor, 17
Visalia Roadster Roundup, 70
Von Dutch, (see Kenneth Howard)
Warunek, John, 14
Watson, Larry, 67
Western Timing Association, 33
Williams, Dick, 55
Willow Springs Raceway, 41
Wineland, Lynn, 40
Woods, Dan, 73, 74, 76, 77